D1473015

America the Powerless

Facing Our Nuclear Energy Dilemma

by Alan E. Waltar

Cartoons by Kathy Kachele

Cogito Books
Madison, Wisconsin

International Standard Book Number: 0-944838-58-8

Library of Congress Card Number: 95-080187

Cover design and cartoon illustrations by Kathy Kachele

For information, address the publisher:
Cogito Books
(Medical Physics Publishing)
4513 Vernon Blvd.
Madison, Wisconsin 53705
608-262-4021

Printed in the United States
Third Printing

To my grandchildren,
and their children's grandchildren,

who have every right
to hold us in the present generation
accountable for turning over a planet
to them worthy of their inheritance.

Acknowledgements

There are literally scores of people who either knowingly or unknowingly have contributed to this book. Numerous colleagues throughout the years have struggled with the frustrations of trying to communicate this difficult yet incredibly important subject to their neighbors, friends, and families in an understandable manner. It is from the many years of sharing in these struggles that I have developed a format I hope will be useful.

A central feature of the current format is the use of cartoons. Though some may think this strange, I can see no reason why this subject has to be so "heavy." I am particularly indebted to Kathy Kachele for her extraordinary skill and selfless dedication of time to create the delightful cartoons. I owe a special debt of gratitude to Steve Critchlow, Tom Woods, and David Erickson, who devoted countless hours from their already hectic schedules to translate my hand-scratched figures into interesting computer graphics. David is to be especially commended for retouching all the graphics into a final, consistent format.

Over the course of several drafts, I have been fortunate to have the reviews of an exceptionally wide group of readers—ranging from scientists and theologians to housewives and high school teachers. A partial list includes James Ahlman, John Allred, Al Amorosi, Chuck Batishko, Stan Berglin, Ann Bisconti, Gwen Bray, Gil Brown, John

Cameron, Gene Cramer, M.J. Crick, Steve Critchlow, Ralph Curran, Jim Deatherage, Gail dePlanque, Marilyn Druby, Milt Eaton, Don Edwards, Richard English, David and Karen Erickson, Mike Fox, Ed Fuller, Karen Gerwin, Martha Glass, Harvey Goldberg, Sue Golladay, John Graham, Glen Graves, Bill Hannum, Ted Hinds, James Hylko, George Jacobsen, Roger Johnson, Kathy Kachele, Richard Kennedy, Mike Kynell, Ralph Lapp, Mike Lawrence, Bill Lee, C. Gordon Lewis, Bob Long, Harold McFarland, Wanda Munn, Pat Murphy, Ken Old, Ron Omberg, Linda Pfenning, Bob Paul, Nikolai Rabotnov, Dixy Lee Ray, Dave Rossin, Petra Scheerle, Glenn Seaborg, George Stanford, Bill Stratton, Bill Terry, Roger Tilbrook, Jim Toscas, Ellen Weaver, Jerry Woodcock, and Tom Woods.

Roger Tilbrook, of Argonne National Laboratory, was unique to this group because of his enthusiasm in providing copies for review to a large group of professionals that I would never have been otherwise able to reach.

In addition to this extensive list, my own family has been very helpful. Anna, my wife, and children Steve, Doug, Karen, and Bruce have all provided excellent contributions (and are still on speaking terms with me!). Julie Gephart, a professional technical editor and close personal friend, was kind enough to edit the entire manuscript. Glenda Schlahta, a professional editor for several well-known national authors, also contributed significantly to several chapters in the final version, both structurally and editorially.

Finally, I would be remiss by not acknowledging the unprecedented personal attention given by the staff of Medical Physics Publishing (MPP). Dr. John Cameron, founder of MPP, provided far more constructive help than would reasonably be expected, and the editorial staff (principally Eileen Healy and Elizabeth Seaman) contributed substantially to the final product.

In the end, of course, I bear the sole responsibility for the advice that was taken. It is my fervent hope and prayer that this final version captures the essence of the excellent advice received.

Contents

Foreword

For decades it has been clear to the scientific community that nuclear energy is destined to play an ever-increasing role for the generation of electricity throughout the world. Yet here in America, the birthplace of commercial nuclear energy, the promise and production of this technology has been brought to a near halt. Why?

Our technology has been very good. Yes, we have made mistakes; and we have learned from those mistakes without injury to the general public. I believe that the stalemate in nuclear energy is the result of poor communication between the scientists and engineers and the community at large. We scientists are, for the most part, poorly skilled in talking in a language the public can understand, and understand poorly the importance of doing so. This has resulted in the proliferation of unfounded fears among the general public and policymakers. The tragic result is that our nation is perilously close to losing a technology that may be our only bridge into a future that can sustain life in harmony with environmental concerns.

It is within this context that I am pleased that this book is being published. Dr. Alan E. Waltar is more than a competent scientist. He is an effective communicator. I believe he has been able to get to the core of the issues in nuclear energy that trouble most Americans. He presents an honest, authoritative, and yet delightfully readable, response to each of these genuine concerns. His holistic view provides a remarkably balanced perspective of how nuclear energy compares to other available energy supplies for the generation of electricity. This book conveys an important message, which will interest and inform every American truly concerned about the condition of our planet and our legacy to the next generation.

Dr. Glenn T. Seaborg

Nobel Laureate, 1951
Chancellor, University of California, Berkeley (1958-1961)
Chairman of the Atomic Energy Commission (1961-1971)

Preface

In the Broadway musical *1776,* Thomas Jefferson, Benjamin Franklin, and John Adams are comically depicted in a heated argument over which bird should symbolize the unifying spirit needed to help forge a new nation, the United States of America. Jefferson argued that it should be the dove, a peaceful bird with grace, beauty, and dignity. Franklin held out for the turkey, a large and imposing bird indigenous to the land; one that provided essential sustenance to our early settlers. Adams thundered his support for the eagle, a majestic and powerful bird; one that could inspire our fighting troops with the images of freedom and strength so desperately needed to win the struggle for independence. Adams prevailed.

It was not until 1872 that the bald eagle officially became our national emblem. However, the dramatic interplay depicted above provides useful imagery of what Americans envision when reflecting on the essence of our country. Like all living organisms, however, even the mighty eagle, which pound-for-pound possesses one of the fiercest fighting forces in existence, is not immune to danger. Enemies lurk all around. Paradoxically, the principal enemy of our prized bald eagle has come from within; the bullets from our own rifles. A moment of rest can be, and has been, fatal to our proudest and most cherished commander of the air. A lone hunter or a band of angry snipers can be deadly effective in imperiling the longevity of this exalted feathered friend.

A similar story can be told of the nation that this proud bird symbolizes. As powerful and successful as our country has become, the energy sources that have fueled our nation to greatness over these past two centuries are gradually being gunned down, mainly from within. Traditional energy supplies are beginning to run out. And newer, potentially inexhaustible replacements are being sabotaged to an alarming extent. Tragically, it is such internal destruction that has initiated the downfall of most great nations throughout the ages. Bluntly stated, the future of our nation, indeed the inheritance of our children, is in serious jeopardy if the "snipers" who are dedicated to opposing new forms of energy are allowed to continue the fear tactics that are paralyzing our nation.

Though few in number, these "snipers" (some even well-intentioned) have been enormously successful in preying on the emotions of an unsuspecting public. They have almost mortally crippled an energy source that has the potential to offer affordable, safe, and environmentally acceptable energy for centuries to come. Vast numbers of our citizens have been led to believe that nuclear energy 1) is not needed, 2) is unsafe, 3) poses an unfathomable waste disposal problem, 4) increases the chances of nuclear war, and 5) represents the ultimate insult to our environment.

This book is not intended for those who intentionally employ scare tactics to serve their own agendas. Rather, it is for responsible Americans who appreciate the quality of life that abundant energy supplies can provide, yet are becoming troubled by environmental questions such as smog, acid rain, and global warming. It is for those who recognize that nuclear energy might represent a solution, but are still genuinely concerned about its "well-known" perils.

The issue of nuclear energy is approached in this book from the perspective of a skeptic, a "Doubting Thomas."

The fears and concerns, deeply felt by many, are recognized as real. Consequently, each chapter is opened with a shaded box containing a paraphrase of a skeptic's view. They form the basis for each chapter. Emotions have been deeply stirred, and it is essential that such feelings be recognized and addressed in a sensitive and meaningful manner.

During the early drafts of this book, I entitled it "While the Eagle Slumbers," because I felt a slumbering eagle would provide useful imagery in symbolizing our country at sleep. My publisher decided the present title would provide even more powerful imagery because, indeed, the lack of a reasoned energy policy will literally leave our nation powerless. But because of the scores of enthusiastic reviewers who loved the eagle cartoons, I have retained them throughout the chapters to help personify our struggling homeland. Every day that we fail to take constructive action, our country is becoming more powerless. Unless we wake her up and take corrective steps, we may one day literally be out of power!

AEW

July 1995

1

SO WHAT'S THE BIG FUSS ALL ABOUT?

. . . A Sobering Look Ahead

A Skeptic's View

Nuclear energy is in trouble. There's no question about it. And who cares? We have plenty of electricity, both now and in the foreseeable future. We don't have to face up to the question of nuclear energy. We can do without it.

It's common knowledge that we're in the midst of an energy glut. Despite the cyclic rhetoric of impending doom, the lights still come on every time I throw the switch. Besides, there is an ethical issue that needs to be addressed; namely, how much energy should Americans be entitled to use? If there is any sense of fairness, we should be using less, rather than more energy. Conservation is clearly the best approach to addressing our energy problems.

> *So why would anybody in their right mind support an industry that manufactures a product we don't need, uses a process that could kill thousands of innocent people in the event of a mishap, and leaves a waste problem that will linger for tens of thousands of years? Isn't pollution getting bad enough as it is?*

Sounds like pretty heavy stuff. Certainly, if the above impressions and accusations were true, there would be much better ways to spend our talents and money. But are they true? Is the energy glut real, and is it here to stay? Is there any legitimate <u>need</u> for nuclear energy?

It is true that our country has gone through a period of plentiful electricity supplies, and the availability of oil has been maintained. Unfortunately, over the past decade, U.S. leaders have been lulled into thinking that it will always stay this way. We've been told that market forces will solve any problems in energy supply and demand. It's not surprising that most Americans, who have plenty of other issues to demand their attention, simply don't recognize the looming danger.

The question of "our fair share" is a sobering thought. As world population continues to soar, and modern communication devices constantly shrink our globe, the issue of equitable distribution of resources deserves careful reflection. However, much of the debate seems to be based on carving up a constant (or finite) energy pie, rather than looking toward sustainable long-term supplies (i.e., a larger pie).

How about the issues of safety, waste, and environmental pollution? Let's put these on hold for a bit. We'll return to them <u>after</u> we've determined whether the whole topic is worth our time.

To make the latter assessment, it is helpful to step back for a moment in history. Let's take a look at how energy use patterns have developed over the past century, both globally and within the United States.

A WORLD ENERGY PERSPECTIVE

If we consider the progress of civilization, we recognize the constant striving to amplify human muscle power for performing difficult tasks. Animal power has, of course, been used for some of these purposes as far back as recorded history. Within the last century, however, the introduction of laborsaving mechanical devices has completely transformed society, particularly in the Western nations.

This is illustrated in figure 1, in which both rising world energy use and world population are plotted. It is clear from this figure that energy consumption is increasing considerably faster than the total population. Per capita energy consumption has risen continuously. There is certainly nothing in this global energy perspective that suggests the demand for energy will diminish.

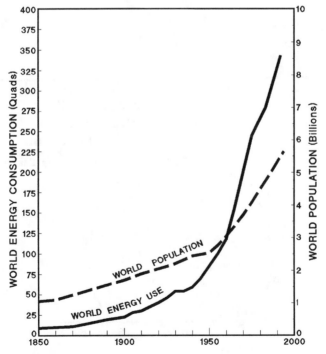

Figure 1. World Energy Use Increasing Faster Than Population[1]

Perhaps a more fundamental question is why the energy demand continues to rise. Figure 2 may be useful in this regard. Gross National Product (GNP)[2] is used as a measure of the total productivity of any given nation. Although GNP per capita does not directly measure the quality of life, most sociologists would concur that the economic health of a nation provides a meaningful index to its overall vitality.

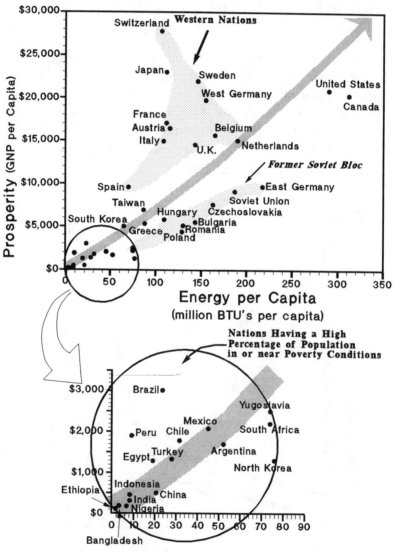

Figure 2. Energy per Capita versus Gross National Product[3]

The curve in figure 2 provides a relative measure of how efficiently any particular nation is able to convert energy usage into productivity (valuable products). Spirited arguments can certainly be made regarding the relative position of specific countries in this figure; for example, considerations of geography, political climate, principal manufactured products, value systems, etc. However, it is easy to recognize the unmistakable correlation between lifestyle and energy usage, particularly if we look at the cluster of nations at the opposite extremities of the curve. One interesting observation is that the former Soviet Bloc countries are significantly less efficient in providing food and goods than the Western nations. This fact may have been the root of much of the unrest that led to the dissolution of the Soviet Union.

Although one may believe that people in the wealthy nations must find a way to curb their seemingly insatiable appetite for energy consumption, it must be recognized that this will not be easy. Even if the Western nations (the U.S. in particular) should consciously elect to exercise restraint, it would be naive to believe that the developing nations (those clustered near the bottom of figure 2) would be satisfied to remain mired in their current living conditions. The ethics of imposing such an energy hiatus on developing countries is, in my judgment, unconscionable.

A UNITED STATES ENERGY PERSPECTIVE

So what does the energy picture look like closer to home? Figure 3 illustrates U.S. energy and population growth since the days of Abraham Lincoln. We see a striking similarity to the world energy use, previously shown. U.S. population growth is somewhat less, although immigration rates continue to offset the declining birth rates. Energy consumption continues to rise at a very strong pace, and energy use is up over the last century by well over a factor of ten.

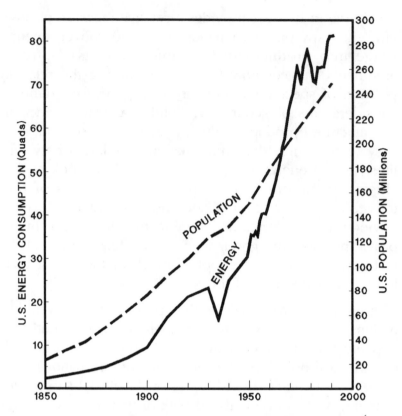

Figure 3. U.S. Energy Use Increasing Much Faster Than Population[4]

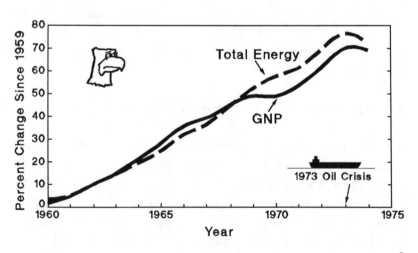

Figure 4. Correlation of U.S. Total Energy Use and GNP Prior to 1973 Oil Embargo[5]

Earlier, I implied that there is a correlation between energy consumption and economic health. This is clearly the case in the United States. Figure 4 illustrates the changes in total energy use and GNP from the early 1960's to the Arab oil embargo of 1973. The almost lock-step correlation would seem to be more than pure coincidence.

During the Carter administration in the late 1970s, major arguments were made against the theory that sustained energy consumption was necessary to fuel the nation's economy. We were told that the solution to our dependency on foreign oil was conservation. Cutbacks could be taken that made both ethical and economic sense, and we could break our strong dependence on energy use.

Americans responded. They began to insulate their homes, turn down the thermostats in winter, and push them up in the summer. Even the automotive industry got the message that gas-guzzling cars were "out."

To be fair, the bulk of this activity was more likely driven by raw economic forces than national concern. As gasoline prices soared, the pressure from foreign auto makers specializing in fuel-efficient cars (principally Japan) was much stronger in Detroit than any appeal from the White House. Likewise, higher heating and air conditioning bills provided more than ample incentive for many Americans to control their thermostat settings.

As we look at actual energy use patterns since the 1973 oil embargo, we encounter some illuminating information. As shown in figure 5, we did indeed break the lock-step correlation between GNP and total energy consumption. In fact, the non-electrical energy use in the U.S. in 1990 was actually less than in 1973. However, the correlation between GNP and electricity use is as strong as ever. Further, the fraction of our total energy usage that is in the form of electricity continues to rise. Electricity furnished

9% of our energy in 1940 compared to 36% in 1990, a growth in importance of 400% in 50 years.

Americans are hooked on electricity.

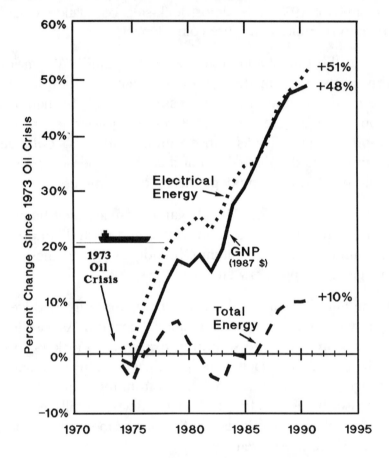

Figure 5. Correlation Remains between GNP and Electrical Energy Use[6]

Stop for a moment and look around. It is almost impossible to find an area of human endeavor that is not supported in some way by the use of electricity. Though we may not be aware of it, the bulk of modern industrial machinery is propelled by electric motors. Many construction materials are melted and refined by electrical heating. Even the transportation sector is beginning to be propelled by electricity (e.g., electrified trains, especially in the

Northeastern states). Computers, refrigerators, air condi-
tioners, and television are so commonplace that a few
hours without electricity makes us feel as if we are living
in a cave. Even oil or gas fired heating systems won't work
when the electricity goes off.

The message is unmistakable. Electricity is becoming more
and more essential to sustaining life in the United States,
as well as in other industrialized nations.

ENERGY AND THE QUALITY OF LIFE

While the historical data unequivocally underscore the
strong link between electrical energy use and economic
vitality, there is still the question of whether an ever-
increasing GNP is necessarily desirable. Many ask if it is
fair for the wealthy Western nations to consume the energy resources of the globe, particularly if it further disad-
vantages the developing nations. This is, in my view, a
disturbing question—sufficiently disturbing that I believe
it warrants more than passing recognition. In fact, a search
for answers to this question provided the principal moti-
vating force for my investigation of the global energy situation. Consequently, much more attention will be devoted
to this point in chapter 10.

For now, however, let's reflect on what life is really like in
the nations shown in figure 2. While volumes could be
written on this subject, most of us recognize there are
some very different realities facing a child depending on
the country in which he or she is raised.

The average life expectancy for those living in countries at
the low energy, low GNP range of figure 2 is 64 years. By
comparison, the life expectancy for the remaining countries (high energy and high GNP) is 76, a full 12 years
longer.

The Overseas Development Council has devised a Physical
Quality of Life Index (PQLI) that uses three criteria to measure how well nations are meeting basic needs for food,
sanitation, medical care, and education. These are life

expectancy, infant mortality, and literacy.[7] Figure 6 uses these factors, along with per capita GNP and energy use, to compare the United States and India. In addition to the marked contrasts in the PQLI, we note that the standard of living in India is so low that citizens spend five times more of their total income on food (bare subsistence) than those of us in the United States. It is difficult to argue the virtues of energy deprivation when pondering these data.

		U.S.	India
Per Capita GNP ($)	(1989)	$20,910	$321
Income Spent on Food (%)	(1988)	10	51
Life Expectancy (yrs)	(1991)	76	57
Infant Mortality (deaths per 1000 live births)	(1991)	10	84
Literacy (%)	(1991)	98	52
Electricity Consumption Per Capita (kWh/person)	(1990)	12,227	338

Figure 6. Improved Quality of Life Afforded by Greater Energy Availability [8]

The American farmer and the entire food production system have allowed those of us living in the United States to devote a substantial share of our income to activities of choice rather than the basics of simply staying alive. High literacy and low infant mortality are advantages we take for granted, but, like the ample and inexpensive flow of food supplies, they can happen only within a system that is supported by a well-fueled infrastructure.

High-quality medical care is another manifestation of what we value in America. Even though we grumble about high medical costs, there are few of us who would choose to take an ailing loved one outside our country for treatment. Modern medicine is one of the major reasons life expectancy continues to climb in the Western world, and in the U.S. in particular. Take electricity away from the pharmaceutical suppliers or the hospitals of America and see how long such care could be sustained.

My former mentor, Dr. Bill Roake, was asked by his heart bypass team if he had any last questions before surgery. After looking around and marveling at the myriad of electronic and mechanical contraptions surrounding his bed, Bill paused and whispered, "And what is your backup system in the event of power failure?"

That is a question we all might ponder.

PROJECTIONS

So where do we go from here?

While projections of any type are difficult and far from precise, it should now be clear to us that it is essential for world citizens to have some basis for knowing what energy options are available to sustain life on this planet. Beyond the basics of simply staying alive, there is a growing concern that such future energy supplies must be environmentally acceptable. If we are truly concerned about the quality of life that may be in store for our children and grandchildren, we must be willing to leave them with energy options that sustain life.

Numerous studies have been conducted that provide roadmaps for responsible energy planning. Perhaps the most comprehensive global energy study yet undertaken was completed in 1980 by the Energy Systems Program Group of the International Institute for Applied Systems Analysis.[9] This group, composed of approximately 150 energy experts from around the world, spent seven years assessing viable paths to a sustainable energy future. They concluded that such a future could be achieved, but only if massive investments were made to wean world citizens away from fossil fuels and into nuclear and solar systems. They also concluded that some systems, probably hydrogen-based, would be needed to replace oil in the transportation sector. Their major plea was that all possible energy sources be maintained and developed, for they will all be needed to create a bridge to a long-term nuclear and solar-based sustainable system.

A more gripping study, recently concluded by Dr. Chauncey Starr of the Electric Power Research Institute (EPRI), focused specifically on the amount of electricity that will be needed over the next 75 years, the expected life span of a newborn child.[10] As former Dean of Engineering at UCLA and President Emeritus of EPRI, Dr. Starr is in a uniquely credible position to probe the realities of world electricity supply futures.

Based on present trends (shown in figure 1), Dr. Starr found that by the year 2060 the world-wide GNP would reach a level 2.8 times that of 1986. For his Case 1 scenario, he determined that the total energy and electrical energy consumption required to support this production level would rise by factors of 4.4 and 7.9, respectively. That is, world electricity generation would have to increase by nearly a factor of eight in 75 years (an annual growth rate of 2.8%).

Recognizing both the ethical and economic advantages of conservation measures, Dr. Starr reworked his calculations by granting full credit to all conservation measures capable of reducing energy consumption without lowering the GNP (certainly a most optimistic scenario). For this Case 2 scenario, the total energy and electrical energy consumption growth factors could be reduced to 2.5 and 4.7, respectively (an annual growth rate for electricity of 2.1%).

For perspective, a Case 3 was developed that was based on zero energy growth per capita, i.e., a complete freeze on per capita energy usage at present values. This yielded total energy and electrical energy increases of only a factor of 1.6 (an annual growth rate of 0.6%), a number reflecting only the 60% population growth. The fact that this case reduced the worldwide GNP to 70% of 1986 levels is but one indication of the political and sociological unacceptability of such a scenario. Furthermore, even this case would still require a substantial power plant building program. It would need to account for the 60% population expansion plus replacing existing electrical power plants, which are typically designed to operate for only 40 to 50 years.

The full impact of this study begins to hit home when we ponder the capability of known means to provide the large amounts of electrical energy that will be required. Dr. Starr considered three principal combinations of energy sources that might have a chance: fossil fuel and hydro-electric, solar and biomass, and nuclear.[11] Starting with 1986 numbers for the world-wide production of electricity, the year 2060 requirements to sustain Case 2 (i.e., maintain a healthy economy with maximum conservation measures) revealed that massive increases in both solar/biomass and nuclear systems would be necessary. This is true even if the fossil/hydro supplies doubled, which is doubtful if environmental efforts to curb fossil fuel burning continue to be successful. The solar/biomass calculation assumed a fully mature solar energy conversion system (with both technological and economic breakthroughs few scientists currently consider even remotely feasible). Even with this unrealistic expectation for the solar/biomass contribution, Dr. Starr found that nuclear energy would have to increase by **over a factor of ten** to fill in the remaining gap.

Going from a world electrical consumption of 10 trillion kilowatt hours in 1986 to 47 trillion kilowatt hours in the year 2060 means that roughly 75 power plants of 1000 megawatts each (San Francisco uses about 1000 MW) would have to be built *each* year for the next 75 years. And, this doesn't include replacement plants. Certainly this would not have to be a constant rate of buildup. These numbers are startling, however, particularly when we consider that the United States has completed only about 50 electrical power plants of this size in the last 13 years (less than an average of 4 plants per year).

The only new orders on the books are for a few small gas-fired (fossil fuel) plants.

Since the energy crisis in the late 1970s, making energy projections for the United States has become almost a national pastime for specialists (and self-proclaimed specialists) in the field. The data shown in figure 3 establish the baseline, but the debate over the U.S. energy future continues to rage.

Bloomster and Merrill compiled a listing of essentially all of the electrical energy demand projections published in the U.S. during the last decade.[12] They included 40 different studies, consisting of 89 scenarios. The average electrical growth rate predicted by these 40 studies was 2.7% per year, including a high of 5.8% and a low of -0.8%. The lowest figure comes from a study conducted by the National Audubon Society and was the only one that predicted a negative growth. The credibility of that number, generated in the year 1980, can be readily tested because the actual annual growth rate of electrical consumption during the decade since 1980 was 2.1%. If we discard the two highest and the two lowest of the projections, the average stays at 2.7%, and the remaining high and low projections are 4.5% and 1.5%, respectively.

A recent national study was based on an expected 2.5% per year growth of electricity use in the United States between the years 1990 and 2010.[13] The study noted that this would require the construction of 550 new 1000-megawatt power plants in the U.S. during this 20-year period. Even if the lowest growth case were to occur (with massively successful efforts in conservation), the resulting 1.3% per year growth projection would still require the construction of 250 new 1000 megawatt plants by the year 2010. The actual growth in U.S. electrical consumption over the decade of the 1980s was an average annual rate of 2.1%, and it averaged 2.6% in the final half of the decade (after we recovered from the recession of the early Reagan years). Unless we choose to engage in a self-inflicted, long-term recession, a substantial new building program for electrical power plants is urgently needed.

My own area of the country, the Pacific Northwest, has traditionally enjoyed the lowest electrical power rates in the nation. The principal reason for this is that a couple of generations ago a number of hydroelectric dams were built along the Columbia River. History records a few vocal opponents who denounced the construction of these dams on the basis that there was no way human beings could ever use all of the electricity that the dams were capable of producing.

However, even here in the Pacific Northwest, with our huge supplies of hydroelectric power, blackouts are not impossible. The Northwest Power Planning Council has consistently stressed conservation measures as the solution to long-term energy problems. Yet, during the summer of 1988, when record heat waves scorched the nation, the "power bastion of the country" was out of power. Bonneville Power Administration officials appealed to British Columbia Hydro in neighboring Canada, and agreements were made to purchase 800 megawatts of electricity for several days (about enough to power the city of Seattle) to prevent a blackout. During the following winter, an arctic cold front swept the northern part of the continent and dropped temperatures to well below zero for over a week. A second panic call to B.C. Hydro was met with the response, "Do you know how cold it is up here?" They simply didn't have any excess power available. Fortunately, California came to the rescue and provided the needed kilowatts (mostly from oil- and gas-fired generators).

During that same hot summer of 1988, Commonwealth Edison, serving the Chicago vicinity, averted a disaster that few people even knew was in the making. On August 2, Commonwealth Edison delivered a record peak of electrical power to meet the burgeoning demand of its customers, barely enough to keep the electrical-grid system from collapsing. A key part of that load was produced by the Braidwood nuclear power plant. I mention this because only three years earlier, in 1985, the Illinois Public Utility Commission nearly prevented the completion of this plant by denying Commonwealth Edison the authority to help finance construction of the plant through an increase in the price of electricity. Their grounds for denial were that the power for this plant would not be needed until at least the year 2005! Yet without the Braidwood plant, the Chicago vicinity would have been faced with rolling blackouts, or total system failure and a blackout throughout much of the Midwest.

On August 5 of that hot summer of 1988, I learned from Richard Wilson, professor of physics, that Harvard University was closed the previous Friday for only the third time in its 350-year history. The reason: no electricity.

The power glut is over. No matter how we look at it, we must face up to the question of providing new means to generate electricity. The question is how can we do it?

2

BUT WHAT ABOUT THE
THINGS I HEAR?

. . . Influence of the Mass Media

A Skeptic's View

Well! If there is such an impending need for new sources of electricity, why haven't I heard about it before? This certainly isn't the impression I get from what I read and hear.

I do my share of browsing newspapers, listening to the radio, and watching the evening news on TV. I don't get the sense of any pressing need for nuclear energy. In fact, it is just the opposite! I've heard about a lot of accidents in the nuclear business. The whole topic is so controversial that even the scientists themselves seem to be split down the middle regarding its virtues. Who am I supposed to believe?

It is true that one would not get the perspective offered in the preceding chapter by scanning through newspapers or by surfing the numerous TV channels that jockey for our attention. What *does* the media tell us?

For starters, let's take the case of the nuclear power plant accident at Three Mile Island (TMI) (just outside Harrisburg, PA). On March 28, 1979, citizens across the land were awakened to learn they were in the midst of the "disaster of the decade." A series of human errors and equipment failures had resulted in a partial meltdown of the reactor core.

The electronic media had a field day. I'll never forget watching "CBS News" the evening of the first day following the accident. Walter Cronkite, probably the most credible man in America at the time, ended the telecast by stating, "I don't know about you, but I'm scared. And that's the way it is."

Wow! I had a hard time calming my own family, located nearly 2,000 miles away from the scene of the accident!

The front pages of our national and local newspapers featured the story for nearly a month. Typical headlines were "Taste of Doomsday in Pennsylvania Nuclear Accident" and "Radiation Spreads 10 Miles from A-Plant Mishap Site."

As we know now, there wasn't a public safety disaster at all (see chapter 5). No one was killed or injured. In fact, if radiation exposure was the principal hazard, the news reporters who flew into Harrisburg to cover the story were the ones at greatest risk. They received greater radiation exposure en route to the site than the residents around the plant. (Again, more on this in chapter 5.)

During the month following the TMI accident, a fire in a Missouri nursing home killed 34 people; approximately 200 people perished in a department store fire in Bucharest, Romania; 40 teenagers died in a bus crash in Zamora, Spain; at least 44 people were killed in Texas and Oklahoma tornadoes; and 235 people died in an earthquake in Yugoslavia. Less than two months after the TMI accident, an American Airlines DC10 crashed at O'Hare Airport in Chicago, killing 275 people. Yet most of these stories got only a single day or two of coverage, or were buried in middle sections of the newspaper.

Why? Why did the media continue to rivet themselves on the TMI story, as if the sky was falling, and essentially ignore stories where scores of people actually lost their lives?

To be fair, hesitation on the part of the Nuclear Regulatory Commission to clarify facts contributed to the confusion in the TMI reporting, especially in the early stages. Still, why was an accident at a nuclear power plant reported as a national calamity and other far more devastating events almost forgotten?

SCIENCE AND THE MEDIA

If we step back and view the TMI story in context, we note that this event marked a rather pronounced shift in the way the media has treated matters of science, especially nuclear science. The discovery of nuclear fission stirred excitement within the scientific community (to the extent news could be disseminated 50 years ago), but it didn't make the front page of the *New York Times*.[1] Part of the reason for its "un-newsworthiness" during the 1940s and into the 1950s was that nuclear fission was first exploited for military use. Heavy classification for national security reasons prevented the topic from entering the mass media.

In 1954, when President Eisenhower signed the Atoms for Peace initiative, the media picked up the story. Initially, the thought of "harnessing the atom" brought a wave of euphoria. This was probably due to at least three things. First, it *was* news because the wall of classification had been lifted. Second, it held the promise of abundant supplies of relatively cheap electricity (an idea few would oppose). And third, there was likely an element of pride because the United States held the lead in this technological breakthrough.

News coverage of commercial nuclear energy waned over the next couple of decades, but the news that was aired was generally positive. It is true that some disturbing news began to appear regarding fallout from atomic bomb tests. The concern was possible cancer. Here the nuclear industry itself was likely at least partially responsible for unnecessarily frightening the public. Scientific groups such as the National Council on Radiation Protection (NCRP) and the International Commission on Radiation Protection (ICRP) adopted a very conservative posture by *assuming* that any amount of radiation, no matter how small, might cause cancer. More recent scientific findings, gathered worldwide, have largely allayed these early concerns (see chapter 4).

Still, most of the media releases prior to TMI were reasonably accurate and balanced. So how do we explain the shift in reporting on nuclear matters that has generally prevailed over the past quarter century?

I don't pretend to know all the answers, but I do know that the fundamental driving forces to achieve success in the fields of science and the media are very different. We have a free press in this country, and we wouldn't have it any other way. Yet given our free enterprise system, it is clear that for any segment of the media to stay in business it has to make a profit. This is not intended as a negative. It is simply a statement of fact.

Given the requirement of the media to sell their product, how do they go about doing it? Beat the competition, pure and simple. In the media business, this means selling more newspapers than the *Morning Gazette* across town, or by achieving a higher Nielsen rating than the other guys. Statistics from 1990 indicate that the gain of a single point in the Nielsen rating during prime time TV for the "Big Three" (NBC, ABC, CBS) translated into over $100 million in annual revenue from advertising sales.

Simply put, the media are in the entertainment business.

Any successful media venture must find ways to make its product more appealing than that of its competitors. Careful topic selection, clever packaging, and rapid turn-around are essential ingredients.

Recognizing these elements for staying in business, how well equipped are the media for dealing with a topic such as nuclear energy?

I am indebted to Dr. Dixy Lee Ray, former head of the Atomic Energy Commission and later governor of the state of Washington, for so clearly elucidating the differences between science and the media.

As noted in figure 7, the only common ingredient essential to the success of either endeavor is that they must have a funding source. Although both science and the media require adequate funding, the very methods by which such funding is derived lead to vast differences in the method of operation. Credibility is the hallmark of good science. Without it, projects and careers are abruptly halted. Consequently, a good scientist or engineer takes whatever time is necessary to do the work required to arrive at a defensible and well-documented result. This requires an in-depth technical background and a willingness to subject the final product to time-consuming peer review to gain professional acceptance.

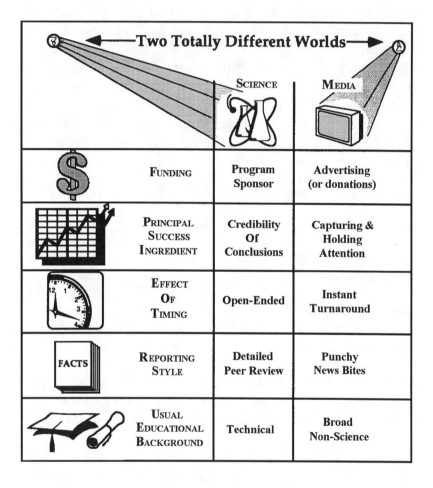

	←—**Two Totally Different Worlds**—→	SCIENCE	MEDIA
FUNDING		Program Sponsor	Advertising (or donations)
PRINCIPAL SUCCESS INGREDIENT		Credibility Of Conclusions	Capturing & Holding Attention
EFFECT OF TIMING		Open-Ended	Instant Turnaround
REPORTING STYLE		Detailed Peer Review	Punchy News Bites
USUAL EDUCATIONAL BACKGROUND		Technical	Broad Non-Science

Figure 7. Science vs. the Media: Two Totally Different Worlds

The media, on the other hand, have very different incentives. Capturing instant attention and holding it are the principal ingredients of success for newspapers, radio, and TV. Although credibility may be important over the long run, the reality is that there is no time to submit news stories to peers for critique and correction. Deadlines are very real. Printed column space or air time is at such a premium that detailed reporting is the exception, not the rule. Watch the evening news tonight and see how much time anyone gets to express a viewpoint on any topic. News bites are typically 20 seconds; almost never more than one minute!

Further, it isn't practical, from an economic standpoint, for any but the largest news organization to have staff reporters with sufficient training to cover specialized news stories adequately, particularly in the world of science.

This is not to say that either science or the media is good or bad. It is simply a recognition that the two worlds are miles apart, and it is no wonder that we often get highly distorted media coverage on scientific matters.

Nuclear energy probably suffers more than any single enterprise in this regard. There is no question that the American public likes sensationalism. We are riveted to the television when we learn of a major disaster somewhere. We line up to buy newspapers and magazines depicting the horrors of a major earthquake or fire. Perhaps it is the ruggedness of our rebellious ancestors that embedded in us our insatiable appetite for action. Whatever the cause, we are hooked on drama.

Given this, what could provide more spice for hungry reporters than to latch onto a problem (even a trivial "hiccup") somewhere within the nuclear industry? The "what if" scenarios are certain to stir up immediate intrigue. Images of a mushroom cloud, hints of a core meltdown, suggestions of lethal releases of radiation (which we can't see, taste, or smell) all are the makings of prominent and frightening news. This is highly saleable press.

Still, why should a reporter gather only selective input, and from only a highly selective group of people, on a scientific issue of significant impact? Worse yet, why report it as if this represented the mainstream of scientific opinion?

A close colleague helped shed light on this seeming paradox for me a few years ago. Paul Lorenzini served as General Manager of Rockwell Hanford Operations (Washington State) for several years in the 1980s before joining Pacific Power and Light Company in Oregon. He earned a Ph.D. in nuclear engineering and later a Juris Doctor in law. Paul explained that it was only after he got his law degree that he understood how this process could occur. The scientific approach is to start with clear evidence and move toward a conclusion. An attorney, on the other hand, starts with a conclusion and moves backward to obtain evidence to support that conclusion.

Will Rogers was right when he mused, "It isn't what we don't know that gives us trouble, it's what we know that ain't so!"

HOW DO SCIENTISTS FEEL?

The sad truth is that on many issues—nuclear energy in particular—the public is often left with the impression that the scientific community is split right down the middle. After all, the two sides are represented on television (or in adjacent news columns) with equal stature. Given such a clear basis for debate, how can the public be expected to make an informed decision on something as important as our energy future if the scientists themselves are equally split?

A poll conducted some ten years ago by Stanley Rothman and Robert Lichter, social scientists from Smith College and Columbia University, respectively, shows that scientists are overwhelmingly in favor of nuclear energy. They used a random sample of scientists listed in *American Men and Women of Science*. Figure 8 contains the results of the 741 replies.

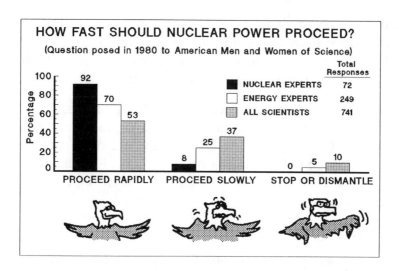

Figure 8. Scientists Strongly Support Nuclear Energy

Rothman and Lichter[2] categorized 249 of the respondents as "energy experts," based on their professional specialties (including atmospheric chemistry, conservation, solar energy, and ecology). They further categorized 72 as "nuclear experts," based upon their specialties in fields ranging from radiation genetics to reactor physics.

From the figure we can readily see that nearly 90% of all scientists felt the United States should continue the pursuit of nuclear power, and over 50% of them felt we should proceed rapidly. Of those scientists with greater knowledge about energy, a full 95% favored nuclear power, and 70% were in favor of proceeding rapidly. For those scientists most familiar with nuclear energy, 100% favored the continued pursuit of the nuclear option, with 92% suggesting rapid deployment. Such statistics hardly support the implied media viewpoint that the scientists are evenly split.

Of course one could argue that nuclear scientists would want to see their livelihood continued and therefore, may be subjecting themselves to a conflict of interest in making such judgements. But, where do we go to seek advice about an erratic heart beat? Do we consult a consumer advocate rather than a heart specialist, because the latter may derive monetary gain by restoring us to good health? Do we seek the services of a social scientist rather than a surgeon in the wake of a ruptured appendix?

Certainly there are "fringe elements" in any profession, and nuclear science is not immune. However, as in any other professional field, most nuclear experts are both conscientious and competent, and continue to pursue their careers because they firmly believe their work is of high humanitarian value.

How Does the Public Feel?

In a democracy it is the will of the people that should ultimately prevail. So what do the American people really think about nuclear energy?

Given the controversy that surrounds this technology, it may come as a surprise to many people that there is strong support for nuclear energy in the U.S. As illustrated in figure 9, a poll conducted by professional groups including the Gallup organization indicates that there is more public support for the use of nuclear energy than at any other time in the decade preceding 1994. Opposition continues to recede.

The February 1994 data indicated 57% were "strongly or somewhat" in favor of nuclear energy, and 37% were opposed.[3] When asked how important a role nuclear energy should play in meeting future energy needs, 71% said they believed it was very important while 26% said it was not too important.

Polls conducted by Dr. Ann Bisconti, a researcher for the former U.S. Council on Energy Awareness, revealed the same basic opinions. Furthermore, her studies revealed that a majority of people in all fields polled (including congressional aides and media personnel) supported nuclear energy. Strangely, none of them think other groups feel the same way.

"Overall, do you strongly favor, somewhat favor, somewhat oppose or strongly oppose the use of nuclear energy as one of the ways to provide electricity for the U.S.?"

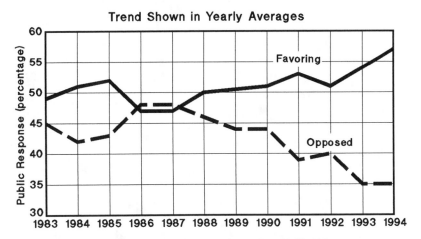

Source: Cambridge Reports/Research Intl., Bruskin/Goldring, and The Gallup Organization

Figure 9. Public Opinion Polls Show Growing U.S. Support for Nuclear Energy

Unfortunately our political and industrial leaders still seem reluctant to take formal steps to revitalize an industry that is almost hopelessly stalled. There have been no new orders for nuclear plants in the United States since before the Three Mile Island accident and, in fact, numerous cancellations have taken place. As we saw in chapter 1, the dangers of not taking action could be very detrimental for both the United States and the developing world.

I believe the main reason for this stalemate is uncertainty and fear. Fanned by the media, a few vocal individuals opposed to nuclear energy have managed to instill us with hesitation. We have been led to believe that we don't need more electricity, and that even if we did, there are safer, less expensive, and more environmentally acceptable sources of energy than nuclear power.

The next chapter is devoted to an assessment of the choices before us.

3

CHOICES, CHOICES, CHOICES

The Electricity Alternatives

A Skeptic's View

Hmmm. So we need more electricity. Maybe so, but we certainly don't have to resort to nuclear energy.

First of all, we haven't even begun to tap conservation. We probably waste more energy than the rest of the world uses! If we would just cinch our belts a bit and be more prudent about how we use electricity, we ought to be able to conserve enough to offset any new electricity needs.

Even if conservation doesn't do the whole job, we have plenty of options to choose from other than nuclear. We've got a lot more oil and gas than the oil barons care to admit. We heard a lot of talk after the 1973 oil embargo that we'd be running out of oil. In reality, the whole thing was a gigantic hoax. The oil companies seem to conspire in orchestrating any "disaster" (such as the Valdez oil tanker spill or the Iraq invasion of Kuwait) into higher prices—and higher profits! How about coal? We're the world leader in coal production. Why not take advantage of that?

Solar energy. Now there's one they can't take away from us. It's free and clean, and it's always going to be there. The only reason the energy sharks haven't developed it is because they can't find a good way to bottle it up and charge us for it! Even wind energy has a lot of potential, it's also free. Then, of course, there is hydro power. That's a particularly attractive one because the lakes are constantly replenished, and they can be used for irrigation and recreation as well.

Anyone who's been to Yellowstone Park and has seen the erupting geysers knows that the earth contains a good bit of thermal energy right beneath its crust. There must be a way to use that so-called geothermal energy. Who says we can't get a lot of energy from burning wood? After the big push for going all-electric in the 1950s and 1960s, Americans finally came to their senses and started going back to good old fashioned wood burning stoves for natural heat.

So why is there any need for nuclear energy? If we have to develop it at all, we certainly ought to stay away from the fission reactors (like Three Mile Island) and go straight for fusion. Who says something like that can't be made to produce lots more electricity?

There is some truth in the above statements. The United States is blessed with an abundance of natural resources. At the time of the 1973 oil embargo the United States imported only about 10% of its total energy supplies, while

Japan imported nearly 90%. Unlike Japan, we have several options.

The problem is that none of these options come without a substantial cost, either in dollars or environmental consequences. Furthermore, most of the familiar ones are not renewable; i.e., the U.S. (and, indeed, the planet) has only a finite supply of conventional fossil fuels. At the present time, approximately 85% of our total energy is supplied from coal, oil, and natural gas, all of which are finite resources. Nearly three quarters of our electricity currently comes from these non-renewable sources.

Somehow we need to wean ourselves away from finite supplies and into a long-term sustainable realm. However, we would like to do it in 'a way that provides a minimum insult to our rapidly eroding environment.

CONSERVATION

As a matter of ethical consciousness, any responsible discussion of energy alternatives must include an honest assessment of conservation. The greater the commitment to conservation, the smaller the need for new energy resources of any kind.

The advantages of conservation are many. First of all, we can do it individually without having to wait for a consensus or a governmental program to get started. We can do it now. In addition to satisfying any moral or ethical imperative that we may personally feel, it also offers a very practical reward by reducing our energy bills.

There is no question in my mind that more, substantially more, can be done in the way of conservation. Technology is now available to provide the same end result using considerably less energy (e.g., more efficient lighting, better insulation to prevent heat loss, etc.). If we have the will to do it, I believe we can stretch our energy supplies well beyond their current efficiency.

Unfortunately, old habits brought on largely by decades of cheap energy are hard to change. While some conservation measures can truly be done for free, such as turning off the lights or reducing the demands for heating or air-conditioning, others require substantial resources for effective implementation. Insulation, for example, requires energy for the manufacturing process. In addition to the cost of good insulation, the proper installation of such materials is (in many cases) prohibitively expensive.

Numerous studies have been devoted to the question of how we can capitalize on conservation. Few people argue with the concept of higher efficiency, getting more useful product for the same energy consumption. Considerably more debate ensues when we consider the question of reducing energy use by either voluntarily or involuntarily changing our lifestyle.

The point is that the total energy savings that we can make by a collective commitment to conservation is limited. Conservation alone will not solve our energy problem, although it can certainly contribute, and should be encouraged. However, from the studies I have seen, even a doubling of our personal efficiency would reduce the national energy needs by only about 10%. Some studies, including Dr. Starr's from chapter 1, suggest that we might be able to achieve conservation energy gains as high as 40%. Unfortunately, even if these very optimistic goals could be met, it would take decades to achieve them.

To put these numbers into perspective, if we sought to reduce our total per capita energy consumption by an exceptionally optimistic factor of two over the next 50 years (time to rebuild energy efficient homes, factories, and office buildings) and if we could immediately reduce our population growth to zero, this would require an average total energy reduction of 1.4% in each of the next 50 years.

This sounds commendable. However, let's see how well we were able to implement conservation measures during the recent period of maximum national conservation consciousness. If we had been able to accomplish an annual reduction of 1.4% in total energy use during each of the 17 years from the oil embargo of 1973 to 1991, we would have achieved a net reduction of 21%. In fact, as figure 5 from chapter 1 shows, our total energy use in the United States over these 17 years of "energy consciousness" *grew* by 10%. To be fair, we did reduce our total per capita energy use by 7% over these 17 years since our population grew by 18% (38 million people) during this same period. However, throughout those years of highest concern following the shock of high oil prices and the massive national efforts to encourage conservation, electrical energy consumption increased substantially—28% in per capita use and 51% in total use. The only period of reduction was during the recession of the early Reagan years.

The message is clear. Conservation, admirable and desirable as it may be, won't solve our problem. If our citizens were not willing to do any better during the time period when energy savings were the easiest to make, we need to be cautious of over-optimistic projections. Any predictions made on the basis of massive conservation gains must take into account the substantial changes in lifestyle that such projections presuppose.

As more and more households include two wage earners, a reduction of the numerous laborsaving devices that we now take for granted will not come without a struggle. Automated heating systems, microwave ovens, refrigeration, and prepackaged foods are part of the reason there is less domestic drudgery than in past generations. All of these are energy intensive. Are we willing to give them up?

OIL

Whether we care to admit it or not, Americans have a special fondness for oil. It currently accounts for approximately half of our total energy consumption. The bulk of it is used for transportation (automobiles, trucks, airplanes, ships, trains), although a sizable fraction is still used for heating and producing electricity. Internal combustion and jet engines have truly revolutionized the world, and until we have a replacement for them, oil will continue to be in high demand.

Oil has many advantages. It is clearly most convenient for transportation. Highway transport vehicles have ranges of several hundred miles with gasoline or diesel fuel, and airplanes can travel up to several thousand miles without refueling. The roads and rail system, now well established in most parts of the world, is fine-tuned in the U.S. and throughout Western cultures. Petroleum is central to most of this infrastructure.

We are only beginning to wake up to the numerous problems associated with oil. To begin with, oil will not last forever. To be sure, we are not in danger of pumping the last drop, but common sense dictates that the chances of discovering new oil fields—especially ones easily accessible—grow dimmer each year. Furthermore, there comes a point where the deteriorating quality of the fields and the expense of drilling combine to require more energy to extract the oil than the energy gained from the product itself. By recording the yield of barrels per foot of drilling, figures for domestic oil fields suggest that the United States may be at the point of diminishing returns within the next decade or two.

Figure 10 shows the principal problem with oil; namely, the location of the world's oil fields. These data, generated in 1990, reveal two important facts. First, North America contains less than 10% of the world's known oil resources (the United States has less than 3%). Second, the Middle East contains nearly two-thirds of the world's known oil resources.

Our appetite for oil is so great that the Arab nations of the Middle East have power far exceeding any military power that they may also possess.

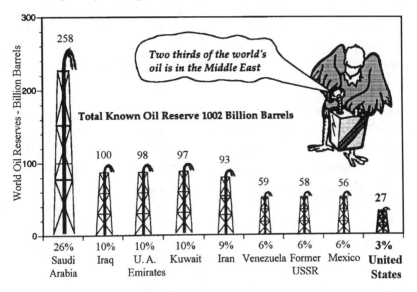

Figure 10. Who's Got the Oil?[1]

Like many others, I have warned for years that our incredible dependence upon these politically unstable nations for energy was akin to tickling the belly of a dragon. Even as early as the Carter Administration, both the Secretary of Defense and the Secretary of State declared that the United States would protect the Persian Gulf at "whatever cost necessary." Make no mistake. The war in the Persian Gulf was about oil.

Certainly there were issues of terrorism and whether a madman should be allowed to bully other people unchecked. Yet how did Iraq's President Saddam Hussein manage to provoke the first war ever initiated by U. S. forces? Oil. Pure and simple.

Of course one could argue about the precision of the numbers included in figure 10. It is true that more oil fields may yet be discovered, and perhaps the world picture could improve in the future. However, it should be noted that the North Sea oil field discovered by the British in 1970 amounts to 30 billion barrels, and the Prudhoe Bay oil field, which feeds the Alaskan pipeline, amounts to about 20 billion barrels. These two massive fields, therefore, amount to a combined total of only about 5% of the world totals. The balance of power for world oil is not likely to change.

How long the world's oil supplies will last is a subject of considerable conjecture. It is revealing to note that the United States consumed 6.1 billion barrels of oil during the Arab oil embargo year of 1973, and 6.2 billion barrels in 1990. The U.S. consumption for the past 18 years since the embargo (1973 to 1991) has averaged 6.5 billion barrels per year. This represents upwards of 1% of the total world supply in a single year. And that is just this country's consumption. World consumption is about three times as large.

Aside from the question of finite supplies, and location of supplies, we should recognize that oil is a fossil fuel. That means that when it burns, it liberates carbon dioxide gas (CO_2). As we will discuss in chapter 9, carbon dioxide is one of the principal ingredients contributing to the so-called greenhouse effect, an environmental problem that may grow to such a magnitude that all previous insults to our ecosystem would pale by comparison. Speaking of contamination, surely we are all painfully aware of the environmental consequences of major oil spills. More on this in chapter 9.

Finally, oil has many more benefits to humankind than its use as fuel. Large numbers of consumer goods are based on petrochemical stocks. Plastics and fertilizers are but two major commodities that require oil as a basis for their existence. Numerous medical supplies owe their origin to oil, as well as do synthetic fibers used in clothing, automobile tires, etc. Given the necessity of using oil to provide such a wide variety of commodities, the burning of this finite resource is beginning to pose serious questions.

About half our oil now comes from abroad, resulting in a trade imbalance of around 60 billion dollars per year. This financial flow out of our country exceeds that for imported automobiles. It is interesting that only the imbalance in the auto industry is causing any significant alarm.

There have been claims that oil could be synthetically manufactured from shale or tar sands. This is indeed possible, and several large scale attempts have been made to exploit this possibility. Unfortunately, this undertaking requires a huge financial investment, results in large waste volumes (at least 1.7 tons of shale and shale residue for every barrel of oil recovered), and requires enormous quantities of water.

In short, we have allowed ourselves to become exceedingly dependent on oil. Its future supply is measured in decades, at best, and the political consequences of maintaining such supplies are now clear. We must preserve this precious resource for its most important uses.

NATURAL GAS

Natural gas has many similarities to oil. It requires relatively simple technology for burning, is fairly cheap (at least at present), and is quite clean compared to most other fuels. Furthermore, it can be used in the transportation sector, although it is considerably less convenient than oil because it has to be contained in pressurized vessels. Because it is a fossil fuel, it likewise generates gases that contribute to the greenhouse effect, although it contributes less CO_2 to the atmosphere than other common fossil fuels. Like oil, natural gas is critically important for many uses beyond consumption as a fuel. For example, it is a major ingredient of the huge ammonia fertilizer industry.

The principal problem with natural gas, like oil, is that of finite supply. Figure 11 indicates that the Middle East contains an appreciably smaller amount of the world's supply of gas than oil, while the former Soviet Bloc has considerably more. Unlike oil, natural gas must be liquefied if it is to be moved by means other than pipelines. Consequently, it is expensive to transport across the ocean.

Although natural gas is often considered one of the safest fuel sources, hardly a year goes by without the occurrence of a major gas explosion. A case in point was the underground detonation on April 28, 1995 (during the final editing of this book), which claimed the lives of approximately 100 members of the public in South Korea. Just the day prior to this accident, a natural gas pipeline exploded in northwest Russia, burning several square miles of forest and sending a fireball some 25,000 feet high.

With an appetite for natural gas similar to that for oil, the United States consumed 18.7 trillion cubic feet of natural gas in 1990. We see from figure 11 that this represents 0.5% of the entire world's natural gas supply in a single year. If we study our domestic production rate for natural gas, along with the expected production rates over the next few decades, we find that upwards of three quarters of our domestic natural gas supply will be exhausted in a single lifetime! Prudence would suggest that we not depend on natural gas as anything more than a temporary fuel.

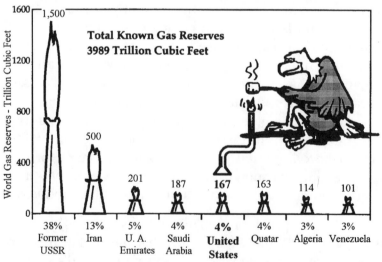

Figure 11. Who's Got the Natural Gas?[1]

Despite this clear lack of longevity, American utilities are beginning to rely more and more on natural gas, largely imported from Canada, as an inexpensive stop-gap measure to meet rising demands for the generation of electricity. It is also gaining appreciable support from the Clinton Administration. Is it really sensible to increase our dependence on yet another nonrenewable fuel, the bulk of which must be drawn from beyond our national borders?

COAL

Coal has been an energy mainstay for the United States, and much of the world, for over a century. Among commonly used fuel supplies, coal is the largest energy resource in the world.

Figure 12 contains an estimate of world coal resources, as compiled by the World Energy Council for 1989.[2] Based upon 1991 world consumption rates (0.9 billion tons in the U.S. and 5.1 billion tons worldwide), it would appear that we may have two to three centuries of useful coal remaining. Furthermore, the United States has a reasonable fraction of the global supply, about 15%.

Coal could contribute to the transportation sector if appropriate technologies can be developed to convert it into a liquid form. It is technically possible to do so. During the late 1970s, there was a substantial push to produce so-called Synfuel.[3] However, several problems made the process ineffective: it is a very expensive process; it requires enormous quantities of water (sometimes a scarce commodity in the vicinity of the coal deposits); and the process produces large quantities of waste, including carcinogenic substances.

Coal has other problems. The bulk of the high energy content coal has already been mined and used. Much of the known remaining deposits are of lower quality, both from an energy content point of view and, perhaps more importantly, from an environmental perspective. The low sulphur-bearing veins are now largely depleted, at least in the United States, and suppliers are being forced to dip into coal deposits that can release large quantities of sulphur dioxide and nitrogen oxides into the stacks.

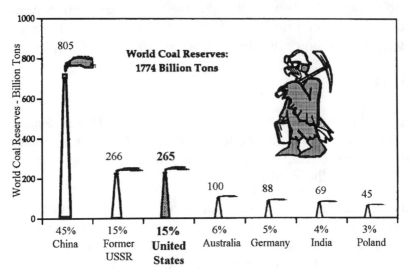

Figure 12. Who's Got the Coal?[4]

To comply with federal air pollution standards, utilities are required to install gas scrubbing devices, which add considerably to the cost of electricity. Even with these expensive scrubbers, sufficient noxious fumes are generated to cause widespread fallout of "acid rain." Already numerous lakes in the United States will no longer support aquatic life. Many believe this is principally due to coal-fired electrical generation plants. Coal plants in the Midwest discharge their gaseous fumes through sufficiently high stacks that they are carried aloft and eventually the prevailing upper tropospheric breezes (approximately eight miles above the earth's surface) transport the fallout beyond our national borders up to Canada. Those who have watched the political interactions between the U.S. and Canada know that the acid rain problem, generated largely by U.S. coal plants, is becoming a very emotional issue in Ottawa.

A few years ago, I had the pleasure of delivering a series of lectures in Karlsruhe, a city located just north of the famed Black Forest of western Germany. My hosts relayed with moist eyes the grief felt by the German people as they witnessed great sections of this revered forest suffer environmental blight. A visit a few years later to that same forest revealed considerably less damage than once feared, yet there are still many who believe the forest remains threatened by acid rain fallout from the massive Ruhr Valley coal burning.

Although the environmental effects of acid rain are still controversial, the detrimental health effects of coal dust in the human lung are undeniable. Black lung disease (a lung inflammation caused by prolonged exposure to coal dust) afflicts thousands of coal industry workers annually. The United States spends in excess of three billion dollars per year to treat the victims of this incurable disease.

As a matter of perspective, coal plants disperse more uranium into the U.S. atmosphere on an annual basis than is used in nuclear power plants. This is because of trace amounts of uranium that naturally occur in all grades of coal. As a consequence, more radiation is released at a coal power station than at a comparably sized nuclear power plant.

The most potentially serious long-term environmental problem associated with the burning of coal is the release of carbon dioxide (CO_2). As mentioned earlier, the buildup of carbon dioxide is widely believed to be the principal cause of the greenhouse effect.

Solid waste represents another concern. For a large, modern coal plant, approximately 100 train cars are required per day to deliver coal. Nearly 10 train cars per day must leave the plant to remove the solid waste.

Still another concern is safety. All of us are aware of coal mine accidents. We grieve when we read of mine explosions or of workers trapped behind collapsed shafts.

Sadly, reports of these tragedies normally don't last on the front pages for more than a day or two because of their familiarity. Yet people continue to die every year in such accidents. On October 5, 1992, a blast in a coal-fired electrical generating station in Indiana killed 3 people and injured 32. The story was printed on page 12 of my local newspaper.

Like oil and natural gas, coal is the basic ingredient in numerous commercial products, ranging from laundry detergents to nylon hosiery. In addition, coal continues to be a key ingredient in the manufacture of iron and steel. Yet over half of the electricity generated in the U.S. is currently supplied by burning coal. Despite its usefulness, we are well advised to ask whether another finite resource, though still available in relatively large quantities, should be burned for the production of electricity.

The United States, along with much of the industrialized world, is blessed with large quantities of coal. But there are some serious safety and environmental questions (specifically discussed in chapter 9) that must be objectively addressed before concluding that coal is the answer to our energy future.

SOLAR

The mere mention of the word "solar" tends to bring a refreshing glow to the energy picture, particularly after we have just waded through the difficulties associated with the conventional resources discussed above. Indeed, solar energy does hold some promise. The possibility of operating a small, self-contained solar electricity unit has considerable appeal to the rugged individualist. The claim that sunshine is free and inexhaustible could be interpreted as valid. Well, to be precise, we might have to admit that the sun is slowly dying out, but the time scale for this is much longer than any of us need to worry about! We need to take a serious look at any energy resource that will last over a thousand years.

Before getting carried away, however, we should quickly tack on a proviso regarding the term "free." Indeed, none of us pay for the sun's rays, but just because the raw source itself is free does not mean that this is the case with the usable end product.

I learned this the hard way a few years ago when I considered going into the bottled water business, as a sideline to my profession. My wife and I discovered a natural spring nestled within a thicket on a small parcel of land we had purchased. Upon testing the water, we were gratified to learn that it was of exceptional quality. Furthermore, the spring was situated in a mountain valley surrounded by a national forest so it would be almost impossible for the water to become polluted from agricultural runoffs. I became excited. I scrutinized the marketing literature and got the confirmation needed – bottled water was a thriving enterprise, exceeded only by wine coolers as the leading area of growth in the bottling industry.

I drew up bottling plans, talked with state inspectors regarding the certifications needed, studied the container, capping, labeling, and boxing supplier companies, and concluded we could make a mint. After all, the water was free. Much to my chagrin, after adding up all the bottling and shipping costs, I was forced to conclude that at the scale of operations I could reasonably expect, I would very likely lose money. The fact that the raw product was free had essentially no value in the entire equation, unless the quality of the water could have drawn a premium price. Fortunately, my level-headed wife pulled me off my "visions of grandeur" ladder just in time to save the house!

Going back to the issue of the sun, it may be helpful to discuss what is meant by the word "solar." If we choose to organize our energy discussions around any system that derives its energy from the sun, we would have to include direct solar heating, electrical generation by both solar furnace and photovoltaic processes, hydroelectric power, wind, ocean thermal energy systems, and even biomass.[5] However, for convenience, we shall deal with these systems individually, recognizing that most people loosely use the word "solar" to connote direct solar radiation, as opposed to those systems that derive their energy as a secondary solar process.

The fundamental problem with solar energy is that it is a very diffuse energy source. If solar energy were a concentrated form of energy, neither you nor I would be able to walk outside. There is no qualitative difference between solar radiation and the radiation generated by a microwave oven. The difference is one of intensity.

During a mid-summer day at high noon in Tucson, Arizona, the maximum solar intensity is about 70 watts per square foot. That might not seem so diffuse, but it is the maximum that can ever exist. It diminishes at other times of the day and year, becomes reduced substantially when clouds form and, of course, is completely absent during the night.

For simple space heating, it is possible in some cases and locations to design systems that make both economic and environmental sense. However, the intrinsically diffuse nature of direct solar radiation suggests it will be very difficult to achieve acceptable economics when attempting to produce reliable electricity. For a plant sized to produce the electrical output of a modern coal or nuclear fired plant, the collection area would have to be approximately ten square miles. This large size translates into the need for a huge amount of materials (concrete, copper, glass, etc.). As a consequence, the cost of solar generated electricity is intrinsically large.

Furthermore, because of the diurnal (day/night) cycle, a complete dependence upon solar energy for the production of electricity would require a massive storage system—

an addition that would be likely to at least double the cost of the electricity delivered. Hence, the most likely role for solar energy, should it become commercially attractive, would be for supplemental power during the day. I know of no responsible energy planner who expects solar energy to be the sole source of electrical supplies of the future, even if consumers are prepared to pay the high costs.

Some may object by saying that it is not essential, nor even desirable to build large electricity stations and the associated interconnected electrical grids. There is a certain attractiveness in the prospect of owning and operating a small solar station, perhaps on the roof of your own home, and becoming completely self sufficient. For such an application, the photovoltaic process may provide some hope. The process is known to work; it is the process employed in several of our spacecraft, where long-term power supplies are needed. Substantial strides have been made in recent years to bring costs down.

It should be noted, however, that the cost will always be high, due to the low intensity of the incoming energy source. Furthermore, it is likely that the initial intrigue of "energy independence" will wear off when it comes time to climb onto the roof to clean the solar panels, repair damaged panels, be responsible for round-the-clock system reliability, and endure the long periods of inclement weather. This is not to diminish the real and growing practicality and desirability of such a system for numerous applications (particularly remote locations), but the reader is challenged to recognize that such a system is unlikely to deliver the enormous quantities of reliable electricity that a growing world population will demand.

A systematic study was done in Canada several years ago to assess the safety of various means of producing electricity.[6] When the results hit the scientific world there was genuine shock. Dr. Herbert Inhaber found that natural gas was the safest all-around system to generate electricity, followed closely by nuclear energy. Approximately 10 times more dangerous were the other methods, oil, coal, and solar. Solar? How could that harmless solar panel resting on the neighbor's roof possibly be more dangerous than nuclear energy or natural gas? The answer to this question is that Dr. Inhaber looked at the total energy balance in determining his results. He started by taking into account normal operation as well as accidents associated with the plants. He also accounted for the total energy cycle in building the plant. When he included the massive amounts of materials that went into building the solar systems, he apportioned the mining and manufacturing hazards associated with the copper used and included the carcinogens that would be generated and distributed around the plants making the necessary glass. This is not to say that a large-scale solar industry would be unacceptably hazardous. Rather, it forces us to recognize that there is no such thing as zero risk.

HYDRO

Hydroelectric power from damming rivers has enjoyed widespread success throughout the world. It has proved to be quite economical and reliable. Hydro power is renewable and a very "clean" way to produce electricity.

As mentioned earlier, those of us living in the Pacific Northwest have a special fondness for hydropower, since the Columbia and Snake River dams continue to produce some of the cheapest electricity in the nation. My last utility bill revealed a rate of 4.2 cents per kilowatt hour. A friend visiting from Long Island, New York, said he was currently paying over 13 cents for that same kilowatt hour.

Hydropower has its problems as well. Perhaps the major impediment to the growth of hydroelectric power is the lack of acceptable sites. Most ideal sites in the U.S. have already been developed, or there are environmental reasons to prevent further disruption of streams and rivers. Although the formation of large lakes can be considered an asset because of the irrigation and recreational opportunities created, very substantial acreage is flooded. This often results in the loss of otherwise productive land. Proposed sites now face years of environmental review and protests.

Even for sites already fully developed, there are rumblings of environmental concern. For example, substantial public pressures have gathered in the Pacific Northwest to increase spillway flow (rather than allowing water to flow through penstocks to generate power) in order to revive the salmon runs.

It is sometimes argued that an advantage of hydropower is that once a dam is built, the system will last essentially forever. While the overall system can, indeed, have an impressive lifetime, this is not always the case. If the river contains a significant amount of silt, this material is deposited in back of the dam, eventually rendering the system ineffective. The huge Aswan Dam in Egypt, which blocks the silty Nile River, may soon be an example of this problem.

It should be noted that large hydro plants built today would be likely to generate electricity costs comparable to fossil fuel plants. The principal reason hydropower is so cheap along the Columbia and Tennessee rivers is that the dams were built decades ago. This was an era of relatively cheap labor and a political climate that lacked most of the red tape that saddles any large construction project in today's world. It is not at all clear that a current replication of those plants, with extensive environmental impact studies and modern construction costs, would be anywhere near as cost effective as we might suppose.

Finally, any system dependent on weather patterns is subject to the question of reliability. Drought, although rare, can and does occur. Any region that relies primarily on hydroelectric power must be prepared to import electricity from non-hydro sources when the rivers run low.

WIND

Ah, if we could only capture those gently flowing breezes. After all, the wind is also "free," and it's inexhaustible.

Well, we've tried. Unfortunately, wind energy suffers from the same fundamental problem as direct solar radiation; the energy density is very diffuse. Although our perception may suggest otherwise, the fact is that wind has only about one-third the energy production capability per acre of surface as direct solar radiation (when practical considerations such as tower spacing are taken into account). As a consequence, enormous land masses are required to produce reasonable amounts of power.

A large wind farm, capable of supplying the electrical output of a modern coal or nuclear fired plant, requires approximately 150 square miles of area. This calculation assumes that the plant operates 40% of the time and that the wind towers are separated by a distance of one-third mile.

Many experiments have been conducted with large wind towers (some with blades 150 feet long and towers 300 feet high), and the experience has been less than favorable. First, it is difficult to find sites where the wind is reasonably consistent. Too soft a breeze results in poor efficiency, and winds too strong have been known to tear the structures apart. Further, objections have been raised concerning the noise generated by the windmills. The low frequency "thud" seems to cause serious irritations for residents living within several miles of the towers. An additional problem associated with this seemingly harmless energy source is that numerous birds, including golden eagles, have been killed by the large rotating blades.

BIOMASS

Wood burning is still the principal fuel in many parts of the world. However, most of us are probably aware that forest denuding promotes serious water runoff problems in many areas and can result in irreversible erosion patterns. Haiti, an impoverished nation not far from the shores of our own country, is a grim reminder of this stark reality.

Still, some have argued that wood could make a significant contribution to our energy futures. To put this argument into perspective, if all the trees in the United States were used for energy production on a renewable basis (i.e., trees would be growing at the same rate they were being harvested), their contribution would amount to only about 10% of our total energy needs. Such an allocation would allow no timber for most of the present uses, such as construction lumber, plywood, wood fiber, paper, etc. This hardly seems like a tradeoff the public would accept. Furthermore, we shall see in chapter 9 that wood burning represents one of the least desirable fuels if we're concerned about environmental pollution.

Other biomass options may be more beneficial. Growing plants to make alcohol for use as a fuel has been suggested. This is not a practical solution because in many cases more energy is required in the total production cycle than is recovered during the burning of the alcohol. It may be of interest for the transportation sector, however, because it is a portable fuel and may represent a substitute for oil.

Special plants are now being developed to produce the energy equivalent of approximately ten barrels of oil per acre per year.[7] If such research is successful, it is possible that biomass could play a significant energy role in the foreseeable future. However, it would require approximately one million square miles of land to produce a quantity of this fuel equivalent to the current annual U.S. consumption of oil. This would consume approximately one-third of the area in the contiguous 48 states. Such land would not be available for any other purpose, and production would be subject to the normal hazards of farming (flooding, drought, etc.). Further, eventual soil exhaustion would occur from one-crop farming.

GEOTHERMAL

It has long been known that molten magma constitutes a major fraction of the Earth's makeup below the outer crust. Consequently, there is an almost unfathomable amount of thermal energy contained in the center of the Earth. In certain regions of the globe, the magma protrudes close enough to the surface of the earth that it meets surface water and creates geysers.

Scientists and engineers have struggled for years to find effective means to tap that buried heat for practical energy production. Some attempts have been successful, but the problems are serious. Probably the most limiting is that there are few places on earth where the hot magma is close enough to the earth's surface to make engineered systems practical. Furthermore, most of these sites are far from major population centers, resulting in high transportation costs for the electricity produced.

Deep drilling is required to reach the best hot areas, and experience indicates that this is a serious problem. Noxious gases almost always accompany such operations. Such gases foul the atmosphere around such a site, and, combined with dissolved minerals, also cause chemical problems with the equipment required for the long-term operation of the plant. The potential for geothermal energy is sufficiently large that further development is probably warranted, but this source is not expected to ever contribute more than a small percent of the total energy needed for an energy-expanding world.

NUCLEAR FUSION

Hopefully, the above discussions of alternate forms of energy production have prepared us to at least consider the nuclear energy option. So, how do we stand on the ultimate form of nuclear energy, namely, nuclear fusion? This form of energy has tantalized scientists for several decades because of the enormous amount of energy potentially available. Nuclear fusion is the process that heats the stars and, in particular, our sun.

In its simplest form, this process involves combining or "fusing" two hydrogen atoms. These are not the ordinary hydrogen atoms that contain only one proton in the nucleus. Rather, they are the heavy isotopes deuterium or tritium, meaning that the nucleus contains one or two neutrons in addition to the proton. Such heavy isotopes, though rare, are found in nature, and the industrial capability exists in many parts of the world to produce them in fairly large quantities. There is enough "heavy water" (deuterium oxide) in the ocean to provide energy for our planet for thousands of years.[8]

The fundamental problem with the fusion process is that the fuel (heavy hydrogen) must be raised to extremely high temperatures, about a hundred million degrees, before the fusion process will occur. At these "interior-of-the-sun" temperatures, the electrons are stripped away from the heavy hydrogen atoms, leaving a fully ionized gas called a "plasma." This plasma, which exists only at these extreme temperatures, must be held together essentially indefinitely if the fusion process is to generate useful energy. Because there are no known structural materials that will withstand such high temperatures, the plasma must be held together by magnetic forces.

This means that the conditions necessary for a functional nuclear fusion power plant are exceptionally difficult to achieve. It is only within the last few years that the "break-even point" has been approached, the point where as much energy is produced as is required to allow the reaction to occur. Such conditions have only been sustained for a few seconds. Even if the plasma can be maintained for the long periods necessary for successful power generation, the metallurgical requirements that must be met by the surrounding structural materials are extremely demanding.

It's precisely because of the enormous complexity of this process that the announcement of a "cold fusion break-through" a few years ago at the University of Utah was met with such excitement. If the process could be achieved at room temperature, the most difficult problem would be solved. Despite initial skepticism among fusion experts, the announcement staggered the scientific community. Unfortunately, it now appears that the claim cannot be substantiated. Despite intense efforts to duplicate the results of the Utah experiments at scientific institutions throughout the world, the results were never repeated. While there will continue to be scientific curiosity about cold fusion, there is now very little hope that such a process will have any positive impact on our energy problems.

This does not mean that hot fusion cannot be made to work. The potential is still there, and some of the best scientists in the world continue to make slow but steady progress. It would appear that numerous safety and environmental advantages may be associated with this process.

Nevertheless, we should be cautioned against placing too much of our hope on nuclear fusion for any near-term solution. Despite the nearly free basic energy source (heavy water from the ocean), the capital cost for such devices is likely to be quite large. Furthermore, all previous experience with shifts to new energy forms (i.e. wood to coal, coal to oil) indicate that a minimum of forty years is required to establish the new infrastructure to allow full scale commercialization. We are not yet even near the starting point for fusion.

NUCLEAR FISSION

So finally we come to nuclear fission. This technology is based on splitting apart very heavy, "fissionable" atoms, thereby producing heat that is used to make steam and, subsequently, electricity. Here we will refer to the fission process as nuclear energy, because it is the only nuclear process that has become commercialized.

Certainly the nuclear energy industry is not without problems. As with any evolving technology, mistakes have been made. The drive to custom design each power plant is an example of the zeal for innovation responsible for driving costs beyond reasonable limits. It might also be argued that this industry has paid so much attention to averting large, highly improbable accidents that it didn't concentrate enough on the small, more likely accidents. With the exception of Chernobyl (to be discussed in chapter 5), an important aspect of the nuclear energy industry is that it has learned from its mistakes without major loss of life.

This safety record is truly remarkable when we consider that nuclear energy currently supplies the United States with 21% of its electrical power supply. Figures 13 and 14 provide an overview of the enormous impact of nuclear energy worldwide. We can see from figure 13 that the United States is the world leader in the total production of electricity from nuclear energy. However, we note from figure 14 that it is low on the list of nations in terms of its commitment to nuclear energy. Fourteen nations now produce a higher percentage of their electricity from nuclear energy than does the United States. France was the international leader in 1990 with a full 75% of its electricity coming from nuclear power plants. Since the dissolution of the former Soviet Union, Lithuania has become the statistical leader, with 87% of its electricity generated by nuclear energy. Even Japan leads the United States in its reliance on nuclear energy. This is especially noteworthy since Japan is the only country in the world to suffer the direct and indirect psychological and physical consequences of nuclear weapons.

What are the advantages of nuclear fission? First, it is a very concentrated form of power, placing far less reliance on fuel supply logistics than the more conventional forms. In its ultimate form, there is an enormous supply—at least a thousand years. Further, despite what the media may have led us to believe, the actual safety record has been substantially better than that of competitive means to generate electricity. The industrial accident rate of the nuclear energy industry is only about one-third that of general industry. Even more dramatic, no member of the public (or power plant worker) has yet been killed by radiation from commercial power plant operations in the Western world.[9] The amount of electricity currently produced by nuclear energy in the United States exceeds the total electrical output of our nation only four decades ago. This is no small achievement.

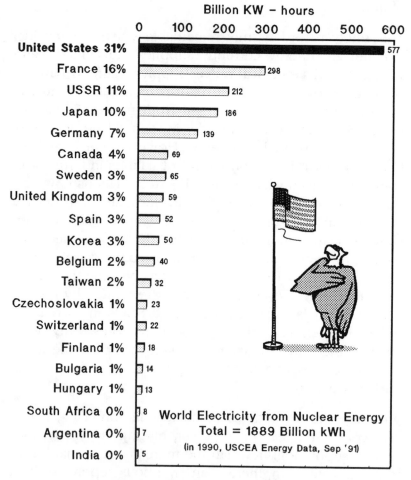

Figure 13. Percentage of the Total World Electricity from Nuclear Energy Generated by Different Countries in 1990 [10]

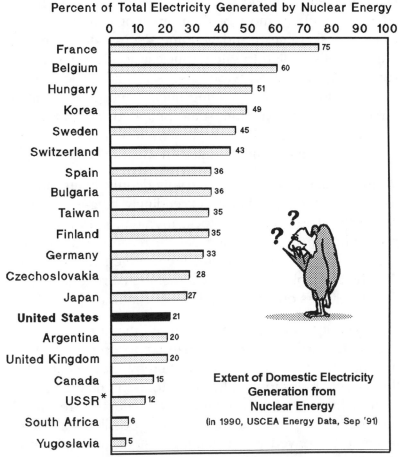

Percent of Total Electricity Generated by Nuclear Energy

Country	Percent
France	75
Belgium	60
Hungary	51
Korea	49
Sweden	45
Switzerland	43
Spain	36
Bulgaria	36
Taiwan	35
Finland	35
Germany	33
Czechoslovakia	28
Japan	27
United States	21
Argentina	20
United Kingdom	20
Canada	15
USSR*	12
South Africa	6
Yugoslavia	5

Extent of Domestic Electricity Generation from Nuclear Energy
(in 1990, USCEA Energy Data, Sep '91)

* Since the dissolution of the former Soviet Union, Russia remains at about 12% nuclear. However, Lithuania became the international statistical leader for commitment to nuclear energy, namely, 87% in 1993.

Figure 14. Percentage of Each Country's Electricity Generated by Nuclear Energy in 1990 [10]

Contrary to what many people believe, perhaps the greatest advantage of nuclear energy lies in the environmental area. The low uranium mining requirements and total absence of atmospheric smoke or chemical discharges during operation make this process very attractive. Also, there is no appreciable competing use for uranium, so its use does not deprive us of essential future needs.

So what are the main difficulties? Unquestionably, the principal problem is public acceptance. Despite the advantages suggested above, most of them are not perceived to be true. The skepticism and genuine deep-seated concerns are generally related to safety (nuclear accidents), radiation (principally high level waste disposal), and nuclear proliferation (the issue of spreading nuclear weapons to different countries). In addition, some people object to the specialized expertise involved, which tends to require centralized control (the so-called "nuclear priesthood").

Further, nuclear energy, if exploited only in its present form, does not represent an exceedingly long-term source of energy. The basic fuel stock, uranium, is in finite supply. Although there is some debate regarding the quantities of available uranium ore, there is general consensus that the available feedstock will fuel the current generation fission reactors only for decades, not centuries.

However, it has long been recognized that it is possible to design fission reactors in a manner to convert "fertile" material into a "fissile" material, thereby greatly extending the useable fuel supply.[11] Uranium found in nature consists of less than 1% fissile material. If a nuclear reactor is designed in a particular manner, it is possible for such a machine to convert the remaining 99% of the fertile (non-fissionable) material into a very useful fissile fuel and at the same time produce large amounts of heat that can be used to generate electricity. This is the fast reactor (often referred to as the breeder reactor).[12] Such a device has been shown to work. In fact, the first nuclear reactor in the world to deliver useful electricity was a fast reactor, located at Idaho Falls, Idaho. The event occurred in December 1951, and since that time the technology has spread to most industrialized nations.

It may be constructive at this point to note that the new fissile fuel that is normally generated from the uranium fuel stock is plutonium. This fact has raised considerable objections to the development of fast reactors from some members of the anti-nuclear community. Their fear is that such a mechanism for generating potentially large amounts of plutonium could exacerbate nuclear proliferation problems (see chapter 7). They correctly point out that a novel feature of fast reactors is that they can be designed to produce more fissile material in the form of plutonium than the uranium that was burned in the process.

Indeed, it is precisely this feature that could change the earth's supply of naturally occurring uranium from an energy supply for a few decades to an energy supply lasting many centuries. For a world that is rapidly depleting its other known energy resources, one would think such a possibility would be viewed as good news.

If there is a legitimate concern with the production of plutonium, it is instructive to point out that a fast reactor can also be designed to be a net consumer of plutonium. Designs are now available that will allow any surplus plutonium to be loaded into a fast reactor as fresh fuel, and the plutonium remaining at the end of the fuel burn cycle is substantially less than the original inventory. Such a process is more difficult to achieve in the "thermal" reactors, i.e., the types of reactors generating the bulk of the world's nuclear electricity today.

Nuclear critics conveniently ignore the fact that current commercial nuclear power reactors routinely generate plutonium as a normal part of the nuclear reaction process. In fact, approximately one-third of all the electricity now produced from commercial nuclear power plants is generated by plutonium. Consequently, if we want to keep plutonium quantities in balance, the fast reactor is a powerful ally. During the times of plentiful plutonium supply (such as the present time, where the end of the cold war has allowed the military to turn over plutonium to the private sector), the fast reactor can consume it while generating useful electricity. At other times when fuel resources run scarce (which will certainly occur at some time in the future), this same machine can be redesigned to generate the quantity of plutonium needed from otherwise useless uranium.

Should the fast reactor be employed to generate plutonium, rather than consume it, the principal new requirement brought about by the breeder reactor is the need to process and recycle the spent (used) fuel. In order to collect the new fissile fuel that is created during reactor operation, the spent fuel must be removed from the reactor and chemically processed. The equipment needed to accomplish this task adds to the cost and to the risks associated with handling highly radioactive materials. Even though new fuel is recovered from this procedure, there is international debate as to whether the value of the new fuel is sufficient to offset processing costs under the current economic climate.

One relatively new development is the possibility of designing an overall fuel cycle/reactor configuration that would allow the reactor to consume (or otherwise destroy) objectionable nuclear reaction products. Specifically, these products are the cause of current controversies regarding the building of a high-level geologic waste repository. More on this in chapter 6.

As a matter of perspective, figure 15 may be useful in visualizing our domestic energy potential available from nuclear energy, as compared to conventional sources. We see that nuclear energy, as generated from uranium by our present power plants, is substantial relative to our oil and natural gas supplies, but it is appreciably less than coal. Should we choose to develop the fast reactor, however, those same uranium resources could be transformed into a fuel supply lasting at least a millennium. Furthermore, there are huge world supplies of thorium that could be used to provide considerably more nuclear energy.

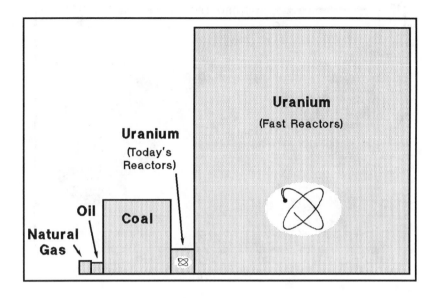

Figure 15. The size of U.S. Energy Resources as of 1980 [13]

ENERGY PERSPECTIVES

In closing this chapter, it may be useful to reflect briefly on the shifting patterns of the U.S. energy dependency over the past 150 years. We see from figure 16 that wood was the principal fuel during Civil War times. We then shifted to coal and then to oil and natural gas. Note that approximately 40 years were required for the infrastructure to shift to a major new energy source. Given the domestic resources depicted in figure 15, it would seem that nuclear energy is well positioned to play a significant role in U.S. energy futures.

As we have seen, all the energy options have promise and all have associated problems. Based upon this review, it should be clear that nuclear energy deserves another look. A perspective on its particular problems, real and perceived, is provided in the next several chapters.

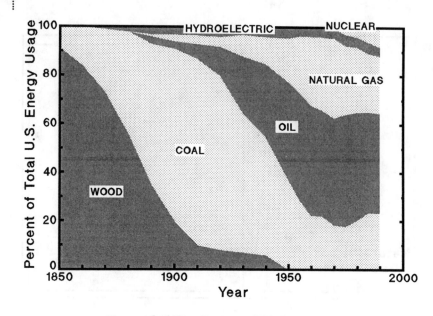

Figure 16. Shifting Patterns of U.S. Energy Use

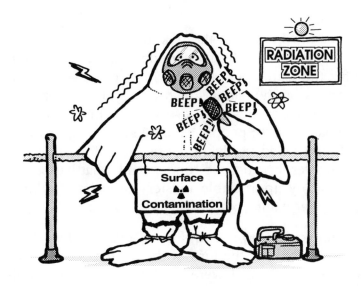

4

RADIATION —

THE INVISIBLE ENIGMA

A Perspective on an Age-Old Presence

A Skeptic's View

Well, it may be true that there are a few problems with our present major energy sources, like oil, natural gas, and coal. Perhaps solar power isn't quite the panacea we would all like it to be. But at least these sources don't give us a radiation problem to contend with!

Radiation is scary. You can't touch it, see it, or taste it. With something that insidious, how can we know if we're being hurt? Of the many man-made technological insults, this one has to be the most frightening.

Most of us have a fear of the unknown. We seem able to accept immense tragedy if we understand what is going on. Fifty thousand people are killed each year in the United States on the nation's highways, yet the number of registered drivers increases every year. In non-nuclear energy related fields, people have been burned, drowned, crushed, and killed by flying debris. Such industries continue, however, because we "understand what happened." If there is any appreciable mystery associated with an enterprise, our natural reaction is to oppose it.

New technologies have always faced this problem. I was told that shortly after the turn of the century, members of the Seattle City Council were considering electrifying the city's streetcars. The council was deeply divided over the issue, mainly because of the safety concerns associated with an unfamiliar element. One council member reportedly called attention to the hazards of the proposition by paralleling electricity to a bolt of lightning. In noting the zig-zag of a typical lightning bolt during an electrical storm, he reasoned that the streetcars would likewise zig-zag off the tracks once electrical power was applied! The records indicate that after much heated debate, the council finally approved the proposal by appealing to the pride of "one-upmanship" over rival city Tacoma.

Nuclear energy undoubtedly suffers more in this regard than preceding technologies. The fundamental reason is radiation. Contrary to fact, many people believe that radiation is new to this century, brought into existence by the nuclear industry. They are surprised to learn that radiation is at least as old as the Earth itself, which scientists now believe is approximately 5 billion years old. The Earth and all of its inhabitants have always been bathed by radioactivity.

A commonly held belief is that radiation was discovered during World War II in connection with the development of nuclear weapons. Actually, radioactivity was discovered in 1896, and both it and its associated effects (both good and bad) have been studied intensively and continuously ever since. Radioactive materials were widely used as portable x-ray sources in World War I and helped field surgeons save many lives. Numerous other diverse applications were started early in this century. The knowledge and study of radioactivity predates the general introduction of the automobile; the invention of radios, airplanes, television, and computers; and the development of rockets, satellites, and space exploration. As in these other fields, our understanding of radiation has continuously advanced since its discovery. Consequently, it is simply not true that "we know nothing about radiation."

So let's take a closer look at radiation; where it comes from, how harmful it really is, and how the advent of nuclear energy has changed our awareness of radiation.

WHAT IS RADIATION?

All matter is composed of atoms, and many atoms are unstable. In fact, over half of the elements in the periodic chart are in a constant process of rearranging themselves. This is not something that humanity had anything to do with; rather, it is simply a consequence of the way the universe was formed.

When the nucleus (the heavy center) of an atom attempts to become more stable, it discharges some type of particle or energy, called radiation. Once this happens, the original atom changes into a new atom, with different nuclear properties. The original atom no longer exists. In some instances, depending on the type of rearrangement within the nucleus, a new element is formed. For such cases, the chemical properties of the new atom are different from the original atom. In other cases, a new form of the original element, called an isotope, appears. Isotopes are materials with the same chemical properties but different nuclear characteristics.

In either event, the time required for 50% of a given radioactive material to disintegrate and become a new element or isotope (i.e., lose one-half of its original level of radioactivity) is known as a half-life. In other words, after one half-life, only 50% of the original material is still around, and after two half-lives only half of the half, or 25%, of the original material is still in existence. The half-lives of various substances differ widely, from very small fractions of a second to billions of years.

We sometimes become concerned when we learn of a radioactive substance with a very long half-life, say a million years, because it sounds like it will be dangerously radioactive forever. Actually, such a substance is likely to be much safer than one with a moderately short half-life, because atoms of the one with shorter half-life are decaying faster, thus emitting their radiation at a faster rate.

A common unit used to describe the intensity of radioactive decay is the curie (named after the French scientist Pierre Curie, who pioneered much of the early work in nuclear science). One curie (Ci) represents 37 billion radioactive emissions per second. Another common unit for radioactive decay is the becquerel (Bq), named after A.H. Becquerel, the scientist who discovered natural radiation in our environment in 1896. One becquerel is defined as one disintegration per second. Accordingly, 37 billion becquerels equal one curie.

As a frame of reference, Cobalt-60 is a highly radioactive substance often used to sterilize medical equipment. It has a half-life of just over 5 years and one pound of this substance contains 500,000 curies of radioactivity. Radium-226, which occurs as trace amounts in the soil and gives rise to radon in homes, has a half-life of 1620 years and contains 454 curies per pound. Uranium-238, the most common form of uranium found in nature, has a very long half-life, 4.5 billion years, but a pound of this material contains only 0.00015 curies of radioactivity.

We often focus on ionizing radiation, because this radiation is sufficiently energetic to knock surface electrons from the atoms of surrounding matter, thus changing their chemical properties. The three most common forms of ionizing radiation are alpha particles, beta particles, and gamma rays.

Alpha particles are identical to a helium nucleus, a fused combination of two protons and two neutrons. An alpha particle is relatively harmless outside of the body. A piece of paper or a layer of skin provides sufficient shielding to prevent penetration. However, alpha particles can be harmful if a material emitting such particles in large quantities is inhaled. Swallowing such material is much less hazardous.

Beta particles are high-speed electrons that are more penetrating than alpha particles. Typically, a thin sheet of aluminum or a similar substance will stop beta radiation.

Gamma radiation is an electromagnetic wave coming from the nucleus of a radioactive atom. It has by far the most penetrating power of the three forms of ionizing radiation. A shield with high mass, such as concrete or lead, is necessary for protection from gamma rays.

X-rays are electromagnetic waves similar in character to gamma rays, but generated by machines.

WHERE DOES RADIATION COME FROM?

As noted above, radiation is an innate part of nature. There is no such thing as a "radiation-free environment." Natural radioactive material exists in the earth, in all food and water, in the air, and in every corner of the known universe, including our own bodies.

Although several units of measurement have been used to determine the levels of radiation exposure, the most common unit is the "millirem," often abbreviated as "mrem." A mrem denotes the effect of radiation on living tissue, regardless of which type of radiation it is or where it comes from.

It is important to remember that the effect of one mrem of radiation is the same, regardless of whether the radiation is natural or created. Many people assume that a mrem of natural radiation is somehow different, or safer, than a mrem from a nuclear power plant. Not so. The human body cannot differentiate.

Figure 17 contains average radiation exposures of a typical American from various sources.[1] These are categorized as natural, artificial, and elective.

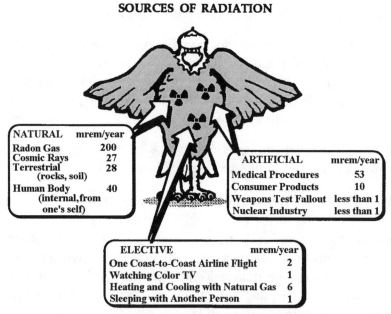

SOURCES OF RADIATION

NATURAL	mrem/year
Radon Gas	200
Cosmic Rays	27
Terrestrial (rocks, soil)	28
Human Body (internal, from one's self)	40

ARTIFICIAL	mrem/year
Medical Procedures	53
Consumer Products	10
Weapons Test Fallout	less than 1
Nuclear Industry	less than 1

ELECTIVE	mrem/year
One Coast-to-Coast Airline Flight	2
Watching Color TV	1
Heating and Cooling with Natural Gas	6
Sleeping with Another Person	1

Total average exposure to radiation is 370 mrem per year

Figure 17. Average Annual Radiation Exposure of a Typical American

We note from this figure that by far the largest amount of radiation is from radon gas. Radon is a naturally radioactive gaseous material that is part of the radioactive decay chain of uranium. This gas continuously seeps out of the ground, plaster and other building materials. The amount of uranium, from which radon originates, varies considerably within the surface materials of the earth's crust, so the amount of radiation exposure from radon gas varies appreciably around the country. The 200 mrem per year attributed to radon in figure 17 represents a country-wide average.[2]

As a point of reference, the U.S. Capitol Building in Washington, D.C., is so radioactive (because of the uranium content in its granite walls) that it could not be licensed as a nuclear power reactor site under today's super-strict nuclear standards. Nor could New York City's Grand Central Station, where employees receive about 120 mrem per year from its granite walls.

A common method of conserving energy is to seal cracks around doors and windows in buildings to reduce air leakage and the associated loss of warm or cold air. By doing so, we trap radon gas inside. This adds appreciable radiation exposure to building inhabitants. This form of conservation is by far the most dangerous energy strategy if we are really concerned about low levels of radiation.

Cosmic rays bathe us constantly from outer space. Again, the number of 27 mrem represents a national yearly average. The principal variable is the altitude where we live, because the atmosphere is a barrier to incoming cosmic rays. The closer we are to sea level, the greater the effectiveness of the atmospheric shield. Annual radiation exposure from cosmic rays increases by about 1 mrem with each 100-foot increase in elevation. Therefore, someone whose home is on a hill 500 feet higher than his neighbor receives about 5 mrem more radiation every year. Inhabitants of the mile-high city of Denver receive about

50 mrem more radiation per year than those in Los Angeles. Airline crews who fly an average of 60 hours per month receive about 120 mrem per year because of the reduced atmospheric shielding of incoming cosmic rays. Some concern is being expressed by airline attendants as they become aware of this fact.

Radiation naturally occurring from rocks, soil, and other surface materials contributes about the same yearly radiation dose as does cosmic radiation. Radiation even comes from within our own bodies. This is mainly the result of Potassium-40 in our bloodstream. It is part of the natural potassium element necessary for sustaining life.

In addition to natural sources of radiation, over which we have no control (except moving to lower elevations and/or locales that have less natural radioactive material on the surface, and avoiding accumulating radon-bearing air in our homes), there are artificial sources of radiation.

The largest radiation source next to "natural background" is from medical procedures. A chest x-ray contributes between 10 and 50 mrem, while a CAT scan may add more than 1000 mrem. Other consumer products, such as luminous dial watches, household smoke detectors, and the older Coleman lantern mantles, contribute much smaller amounts. However, even they contribute over 10 times more radiation than atmospheric fallout from nuclear weapons testing or from the commercial nuclear energy industry. It should be clear from these data that hospitals and other medical facilities, not the nuclear energy industry, are by far the biggest source of man-made radioactivity in the environment.

Finally, I have included a general category of non-medical elective sources. These include the detectable doses of radiation we receive when we choose to fly, watch color TV, or cook with natural gas (the problem here is again entrapped radon gas). Even sleeping with another person poses a radiation "problem," because every person on earth is naturally radioactive (mainly from Potassium-40).

HOW HARMFUL IS IT?

Given that we are all exposed to radiation every day of our lives, what do we know about the health effects to our bodies? Much scientific study has been devoted to this topic. Unfortunately, it is difficult to get a useful perspective on this issue.

The underlying reason for the controversy surrounding this question is that scientists really don't know the precise lowest level of radiation where there is a net harmful effect to the body. I use the word "net" here advisedly, as we will see later.

The best data available on the incidence of leukemia (the most prevalent form of radiation-induced cancer) are associated with Japanese atomic bomb survivors. We know that doses of about 500,000 mrem, delivered over a very short time span, produce leukemia in about 20% of people exposed to such levels. The lowest dose level for which reliable data has produced observable adverse health effects is around 100,000 mrem, where the leukemia rate is about 2%. The question is, what are the dangers of radiation-induced cancer at the even lower levels, say 1000 or 10,000 mrem?

Before addressing this question, we need to remember that very few cancers are actually caused by radiation. In fact, less than 1% of all cancer cases are related to ionizing radiation in any way. Food consumption patterns and smoking account for about 80%. Leukemia is the form of cancer most likely to be induced by radiation, and radiation accounts for less than 10% of the total leukemia incidence. Indeed, at the low radiation levels customarily experienced, radiation-induced leukemia may be essentially zero.

A wealth of data now exists for humans exposed to low levels of radiation. It should be comforting to know that there are no known scientifically documented cases where low levels of radiation (i.e. up to 10,000 mrem) have caused any detrimental health effects. Nonetheless, most health physicists conservatively assume a straight line, or linear hypothesis health effects relationship, as illustrated in figure 18. This theory presumes that any amount of radiation is hazardous, and the amount of damage to the human tissue is directly proportional to the radiation dose level.

It is worth noting that this conservative posture was initially adopted by national and international groups such as the National Council on Radiation Protection (NCRP), the International Commission on Radiation Protection (ICRP), and the National Academy of Sciences Advisory Committee on the Biological Effects of Ionizing Radiation (BEIR). As more and more data became available indicating no observed adverse health effects at low levels of radiation, pressure began to mount within the scientific community to alter this approach. Many scientists are now arguing that unnecessarily restrictive radiation standards are costing American taxpayers huge sums (possibly exceeding $1 billion per year) for safeguards that do not have bearing on public health.

Beyond the linear method, there are three additional ways to project from the known health effects of high dose levels (the Japanese atomic bomb data base) to low levels. Curve A is based on a threshold effect, meaning that it assumes that low levels of radiation cause very little, if any, net damage. Curve B assumes that low levels of radiation can even be beneficial (the hormesis effect), as discussed later in this chapter. Curve C, on the other hand, assumes that low levels of radiation are relatively more harmful, on a per unit basis, than high doses.

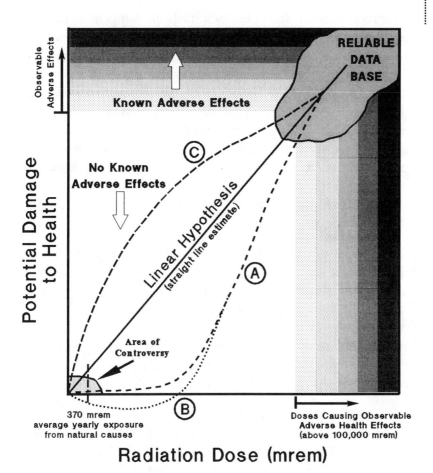

Figure 18. Models for Determining the Health Effects of Radiation Dose

Scientific opinion overwhelmingly favors Curve A. For a detailed and easily understood treatment of the basis for this position, an excellent reference is Bernard Cohen's masterful book *Before It's Too Late*.[3] Dr. Cohen, a professor of physics at the University of Pittsburgh, has devoted most of his distinguished career to the study of radiation and its associated risks. With a publication record of more than 200 articles in scientific journals, he is highly regarded by international health physics professionals. Another excellent source of information is Merril Eisenbud's *Environmental Radioactivity from Natural, Industrial, and Military Sources*.[4]

Numerous studies have led scientists to support a threshold-type damage relationship as depicted by Curve A. These include widely accepted theories on how radiation induces cancer, studies of chromosome damage by radiation in human white blood cells, and experiments with mice and dogs. To be on the safe side, however, scientists responsible for setting international radiation guidelines for industrial safety have deliberately chosen a conservative approach, i.e., to overestimate the risks. They have based such guidelines on the linear hypothesis, the assumption that any level of radiation, no matter how small, contributes to the risk of cancer.

Indeed, it is hard to argue that zero damage is done to the human body by the incidence of a single alpha, beta, or gamma ray. There is a finite probability that a chromosome could be damaged and some specific type of cancer could result. However, all of us are bombarded by an incredible number of such particles and rays every moment of our life (literally some 15,000 particles per second), and we don't all die from cancer. Why?

Our bodies have a sophisticated immune system. Even if transmutations do occur from such radiation (recall that more than 80% of it is "natural"), the body has a way of repairing itself so that no lasting damage is done. There are many substances that can be fatal when taken in large doses, yet are beneficial in small doses. Some vitamins, vaccines and other medications are in this category. Aspirin is a classic example. An overdose of aspirin can kill us. Yet many of us take this "modern miracle drug" routinely in low doses. I do. Current medical literature suggests that one aspirin every day or two can have significant value in warding off heart attacks.

Let's suppose the pharmaceutical community were to adopt the linear hypothesis approach for the health effects associated with taking aspirin. We know that taking 100 aspirins in one sitting would likely cause us considerable harm. In fact, for the purposes of this illustration, let us assume that there is a 100% chance that it would be fatal. Based on the linear hypothesis theory, the practice of taking a single aspirin per day for 100 days would prove fatal. This method doesn't recognize any benefits of time or the possibility that the body may be able to deal constructively with low doses. Yet this is the conservative theory accepted by the radiation industry for the determination of health risk assessments. Even worse, this linear hypothesis method is often applied to an entire population to determine the total "person-rems" associated with a particular radiation release. This was the approach used to predict the number of potential cancer deaths resulting from the Three Mile Island and Chernobyl accidents. Coming back to our aspirin analogy and applying it to a population of 100 people, this method would predict one death if each person took but a single aspirin!

Until now, medical radiation specialists in the Western world have based their studies of the effects of high radiation doses on the Hiroshima and Nagasaki atomic bomb data. However, information is beginning to surface from the nuclear weapons installations in the former Soviet Union that may offer new data. Though preliminary in nature, the data from personnel exposed to doses in the range of 50,000 to 100,000 mrems per year (much higher than any exposures recorded in the U.S. nuclear weapons industry) seem to indicate that there may be a significant rate effect associated with radiation damage. It appears that the human body may be able to survive very large radiation doses if the radiation is delivered over a relatively long time period rather than suddenly, as in the case of a nuclear bomb.

Radiation oncologists (cancer radiation treatment specialists) already recognize and take advantage of this relationship. They are able to deliver more radiation to a malignant tumor with less damage to healthy cells by specifying more treatments with a smaller dose per treatment. If careful scrutiny of this new Russian data can confirm the rate effects, a strong basis would emerge to prove that our current practice of using the linear hypothesis is grossly conservative. The human body may be far more resilient to the effects of ionizing radiation than we have previously assumed.

A recent 10 year investigation of the effects of low level radiation on over 70,000 shipyard workers is most illuminating.[5] It provided perhaps the best epidemiological study conducted to date for cancer and mortality associated with low level radiation. The focus was on three specific groups who received the following work-related gamma radiation doses during the 10 year period: 1) over 500 mrem, 2) less than 500 mrem, and 3) zero. All workers in the study received about 3000 to 4000 additional mrem during this decade of observation due to natural background radiation. The data conclusively showed that both groups of nuclear workers (i.e., groups 1 and 2) had a *lower* death rate from leukemia and lymphatic cancers than the nonnuclear workers (group 3). In fact, those receiving the higher radiation doses had a death rate 24% lower than those who received no radiation beyond normal background.

Perhaps the most dramatic demonstration of the time effect on radiation damage to humans is derived from workers who painted radium on watch dials.[6] Many of the workers received huge doses. Yet there were no reported radiation induced bone cancers for dial painters with less than 22,000 rem to their bones (this is 22 <u>million</u> mrem).

Other evidence is beginning to mount that, as is the case with aspirin, the human body actually benefits from a certain amount of low level radiation. The principle for this assertion is called hormesis, shown as Curve B in figure 18.[7] Dr. T. D. Luckey, a professor of chemistry at the University of Missouri, has presented impressive evidence for such an assertion in his landmark book, *Hormesis With*

Ionizing Radiation.[8] He bases his thesis—that low levels of radiation are actually beneficial—on over 1,000 separate studies with plants and animals.

If true, this hormesis effect could help explain some of the strange phenomena that have baffled epidemiologists (medical experts who study disease control) for decades. The radioactive rock in Kerala, India, for instance, generates a radiation exposure level ten times the U.S. average. Yet residents are reported to have the best health status in all of India, despite only average health care and the least adequate diet in the country. Numerous cases there and elsewhere of life spans routinely over 100 years can be cited, and the one thing these people have in common is that they live in regions with high natural background radiation (e.g., the Hunzan Himalayas in Asia, or the Vilcandra Andes in South America). Here in the United States, Colorado and Wyoming are both well above the national background radiation average, yet are far below the national average in their incidence of cancer. Indeed, of the many epidemiological studies of people living in high background radiation areas, there has never been an instance of a demonstrated increase in cancer.[9]

The above discussion is not to say that any scientific consensus has been achieved on the theory that low levels of radiation might actually be good for us. In fact, this is currently a very controversial area. Rather, it is to suggest that there are several reasons to believe that the linear hypotheses theory depicted in figure 18 represents an overly cautious interpretation of the health hazards of radiation.

There is a very small segment of the scientific community that believes that low levels of radiation are particularly harmful (Curve C of figure 18). However, studies with this conclusion have been thoroughly repudiated by the international scientific community. To me, there is no question that the linear hypothesis represents an overly cautious approach to setting radiation health standards.

It is worth noting that the occupational exposure limit set by the federal government is 5000 mrem per year for people who work with radiation in industry. Using the conservative approaches outlined above, one can see that this level is far below the point for which there is any valid basis to expect detectable injury to occur. In actual practice very few workers receive radiation exposure levels even approaching these limits. The comparable exposure limit for any member of the public is one-fiftieth of this, or 100 mrem per year. Most people are exposed to natural background levels of radiation much higher than this on a continuing basis.

The annual extra radiation exposure for someone living next to a nuclear power plant would only be about 0.08 mrem per year, while the presence of that plant yields only 0.01 mrem per year to the environment as a whole. In other words, the average person would have to live to age 300 to absorb as much radiation from an American nuclear power facility as he or she would receive from a single round-trip airplane flight from New York to Los Angeles.

CAN RADIATION BE BENEFICIAL?

Until this point in our discussion we have generally viewed radiation from a negative perspective. We recognized it as having been with us since the beginning of life, but our goal has been to determine its potential for harmful effects to the human species.

There are, however, direct benefits associated with the use of radiation. Most of us have had x-rays (electromagnetic waves generally less energetic than gamma rays) taken at some point in our life. Chest x-rays are now commonplace for diagnosing lung ailments. Most dental procedures are preceded by x-rays to reveal cavities or unhealthy root conditions invisible to the naked eye. It is a rare orthopedist who performs surgery without first x-raying the joint or bone area involved. More energetic gamma rays are in widespread use for the sterilization of medical equipment.

An enormous array of radionuclides (radioactive atoms) can now be made in nuclear reactors or special particle accelerators. It is possible to use the special "signature" characteristics of such isotopes to perform a wide array of medical functions. Some isotopes are employed to diagnose diseases. Thallium-201 is a radionuclide commonly used to determine the location of blood clots or other blocked passages in the circulatory system. Technetium-99m is commonly used to detect bone cancer and a host of other abnormalities.

In addition to diagnosis, some forms of radionuclides are used as a vital part of therapeutic treatments. Carefully focused radiation is now routinely used to kill or mitigate the spread of cancer cells. A particularly exciting new development is the science of monoclonal antibodies. This process chemically attaches radioisotopes to specific antibodies that seek out and lodge in cancer cells. Although its effectiveness is still somewhat controversial among nuclear medicine specialists, this procedure allows high radiation doses to be delivered directly to the target cells. As a matter of perspective, well over 10 million nuclear medical procedures using radioactivity are performed every year in our nation's hospitals. One out of every three U.S. hospital patients benefits from radiation treatment in one form or another.

The agricultural and food production industries also benefit from radiation. Numerous plant varieties have been developed using radiation. The shelf life of many foods is increased significantly by irradiation techniques. Much of the food consumed by our astronauts is preserved by irradiation. Unfortunately, such techniques have yet to be employed on a mass scale for the American consumer, mainly because of fear associated with the word "radiation," even though tests carried out over many decades clearly reveal no negative health effects. Sad to say, it may take a few more *E. coli* outbreaks from fast food restaurants before the American public allows food irradiation to become commercially viable.

Radiation is commonly used in everyday industrial processes. Because we know the thickness of certain types of materials necessary to shield against (or stop) radiation, it is possible to invert the process and pass a known quantity of radiation through a material of unknown thickness. By recording the amount of radiation that penetrates the material, we can readily determine its thickness. Gauges of this type are now commonplace for measuring and controlling thickness in rolled metals and in a host of other products.

Radiation is used to detect changes in the environment. This is the basic idea of the smoke detector, where smoke passing between a radionuclide and a detector will cause the air ion strength in the detector to drop, thereby setting off an alarm. The radionuclide Americium-241 is the principal ingredient in millions of home smoke detectors.

Other industrial processes use high levels of radiation to modify certain properties of a material. For example, polymer sheets can be irradiated to produce crosslinks that cause the material to shrink when heated. Such "heat-shrink" materials are commonly used as covering for electrical connections and the like. Numerous other examples abound, and the list is growing every day.

Certainly we cannot overlook the potential hazards of radiation, particularly in large amounts. On the other hand, we cannot become paralyzed by a naturally occurring phenomenon that, according to all credible scientific data, has no harmful effect at low levels. Indeed, there is mounting evidence that we may not be able to live without it.

5

AFTER CHERNOBYL? YOU'VE GOT TO BE KIDDING!

A Candid Look at Reactor Safety

A Skeptic's View

So let's suppose that radiation is a natural part of our world. O.K., but this is all low-level radiation. If we could be assured that nuclear reactors were fail-safe and only contributed a small amount to our natural background radiation, perhaps we could live with them.

The problem is that we can't rule out accidents. Accidents will happen. There is enough high-level radiation simmering inside one of those big reactors to wipe out the whole human race. You can't tell me that we can build and operate those complex machines without making a mistake at some point, and when that happens....

Look at Three Mile Island. We were told that an accident like that couldn't happen. But it did. Right here in our own country, a short distance from a major population center, a core meltdown occurred.

Just when we thought the industry might be learning something from its mistakes, along came Chernobyl. That was horrible. It killed lots of people, contaminated vast areas of land, perhaps leaving it uninhabitable forever, and sent radioactive fallout literally all over the globe. From what I hear, several thousand people will die of cancer as a result of Chernobyl fallout.

I might be able to accept nuclear energy if I really believed we could never have an accident. The fact is that accidents will happen; they already have and they will again. We simply can't tolerate uncontrolled releases of deadly radiation.

Indeed, a major accident in a nuclear power plant *is* something we need to be very concerned about. After only a few days of operation, the core of a nuclear reactor is highly radioactive. Following a year of full operation, a large central station power reactor contains approximately ten billion curies of radioactivity, and (as we recall from chapter 4) each curie represents 37 billion disintegrations per second. If it were possible for all this material to become vaporized and be distributed into the atmosphere, it could potentially kill several thousand people.

Before scrambling for the nearest bomb shelter, however, we need to ask the questions: can it get out?; where will it go?, and what will happen if it does get out?

THE BASICS OF REACTOR SAFETY

From the inception of the nuclear industry, nuclear scientists and engineers have focused a good deal of attention on safety. The fact is that the attention given to safety in

After Chernobyl? You've Got To Be Kidding!

81

the nuclear industry far exceeds that of any other major industrial endeavor.

This is not just rhetoric. Companies, universities, and laboratories responsible for the design and construction of reactors have both personal and institutional reputations at stake. They are not about to take short cuts where a real safety issue exists. The Nuclear Regulatory Commission (NRC) is heavily involved in the regulation of reactors built within the commercial sector. Anyone familiar with this agency knows that the safety standards set, the licensing procedures established, and the enforcement practices followed are the most rigorous and exacting within the federal government. All commercially licensed reactors in the United States are monitored by the NRC every day of the year.

For those who distrust government agencies, there are provisions for the public to become directly involved in the licensing process. All commercially licensed plants have been carefully evaluated by both the best engineering professionals and interested members of the public before an operating license is issued. The whole process is intended to ensure that the plant workers and members of the general public are not subjected to any undue risk.

How well does the process work? Based on experience, I believe any objective assessment would have to conclude that the process works very well indeed.[1] The fact is that the nuclear energy industry has an exemplary safety record. Compared with other large industries, the number of lost workdays is one-third, occupational illness is one-fourth, and the number of work fatalities is half. These figures correspond to non-radiation induced hazards; i.e., they relate to any industry that involves constructing and operating large machinery.

Perhaps the most serious concern is with radiation. How many members of the public have died as a result of radiation release from the commercial nuclear power industry in the Western world? None.

None? That's right, none. There have been a few workers who died from excessive radiation within the military and the national laboratories during the early years following the discovery of nuclear fission. Likewise, there were reactor operators and firefighters who died in the aftermath of the Chernobyl accident. The fact remains that no member of the public in any part of the Western world (including the United States, Canada, and Western Europe) has died of radiation-induced injuries resulting from commercial nuclear power. This is the safety record of an industry that now provides 21% of all the electricity produced in the United States. As a matter of perspective, this power block is equivalent to over a dozen Grand Coulee Dams—more electric power than the total used by the U.S. at the close of World War II.

Do these statistics mean that we'll never have public fatalities resulting from radiation released during a nuclear power plant accident? Certainly not. No such guarantees can be made in any industry. Rather, they should provide at least some degree of comfort for those alarmed by the possibility of accidents. A nuclear power plant cannot blow up like a bomb. It is a physical impossibility. Special fuel must be carefully machined and imploded at lightning speed into precise configurations to cause a nuclear detonation. With all the special geometric features designed into a power reactor to allow for steady-state heat extraction, fuel reloading and the like, power plants don't even come close to resembling a nuclear weapon. They are physically incapable of behaving like one.

Given the actual safety performance of the nuclear energy industry, why is there so much fear? Dr. Robert duPont, a psychologist who has studied this problem in some detail, has written a book on this topic called *Irrational Fear*.[2] He concludes there are three basic factors that lead people to develop fear: 1) lack of control, 2) catastrophic events, and 3) unfamiliarity. We seem to fear most those things over

After Chernobyl? You've Got To Be Kidding!

83

which we have no control. Also, we are willing to accept tragedy if spread out over time, but we abhor large events that take great numbers of human lives suddenly (i.e., volcanic eruptions, earthquakes, explosions). Above all else, we fear the unknown.

Dr. duPont believes nuclear energy suffers from public lack of confidence because it combines, at least to some extent, portions of all three elements. Most people feel they have no control over what happens at a nuclear plant, they fear an accident could inflict serious health hazards on great numbers of people, and they have only a vague idea of how the process works.

THE BARRIERS INVOLVED

Since the dawn of the nuclear age, responsible nuclear power plant design engineers have used multiple barriers to ensure that the highly radioactive material within the core of the operating reactor can never escape to the surrounding atmosphere in an uncontrolled fashion. Figure 19 illustrates the four fundamental barriers.

The first barrier is the fuel material itself. When the fission process occurs, the original fission atom (usually an isotope of uranium) splits into two roughly equal parts (called fission products), yielding a large amount of energy. Most of this released energy is carried off by the two newly created fission products, speeding away in opposite directions. In the process of being slowed down and stopped by the surrounding fuel, these fission products heat up the fuel. Cooling water (or some other coolant) carries the heat to a heat exchanger, where the generated steam drives a turbine that generates electricity.

The fission products represent the greatest radiation hazard in the entire process. Most of these new particles are solids, rather than gases (none are liquids), and lodge within the fuel only a small fraction of an inch from where they were born. Those fission products that are gases (such as xenon) eventually percolate through the fuel material and collect in a space provided at the end of the fuel column. The point here is that most of the highly radioactive material is locked up within the fuel and goes no further.

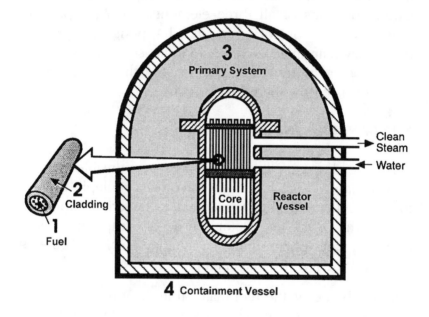

Figure 19. Physical Barriers to Prevent Radioactive Release from a Nuclear Power Plant

The second physical barrier is the cladding, which is a metallic tube surrounding the fuel column. One purpose of this cladding is to provide structural support for the fuel during loading into the reactor. A second very important function is to prevent radioactive fission products (especially the gas) from getting into the coolant. The coolant flows on the outside of the cladding walls to carry away the heat.

Next is the primary coolant boundary. This boundary consists of a heavy steel pressure vessel (usually over six

After Chernobyl? You've Got To Be Kidding!

85

inches thick) and the associated heavy-wall steel piping that carries the hot coolant to the heat exchangers (or steam generators) within its own closed system.

Finally, all commercial nuclear power plants licensed in the Western world have an outer containment system. While there are several designs for such systems, a typical installation may consist of a heavy steel-reinforced concrete structure (perhaps four feet thick), lined with an air-tight steel plate approximately one-half inch thick.

Under normal operating conditions, the highly radioactive fission products never get beyond the second barrier, the cladding jacket. In fact, most of them never get out of the fuel (the first barrier). It is only in the case of an accident that a concern arises. If the power level of the reactor should suddenly rise, or if the coolant should be lost, the fuel temperatures would rise, and the cladding could either melt or be ruptured by pressures generated inside the fuel. Should that happen, radioactive debris could get into the coolant stream.

In order for such debris to get out of the reactor system, there would have to be a break in the system piping or reactor vessel (the third barrier). Such a break is highly unlikely, given the mammoth sizes of the primary boundary structures. To protect against such low probability events, however, the containment system (fourth barrier) has become mandated. These containment systems have enormous strength. As an example, the containment building for the Seabrook reactor, in New Hampshire, is designed to withstand a direct crash of a U.S. Air Force FB-111A bomber, a 360 mile-per-hour tornado, and an earthquake measuring 6.5 on the Richter scale. This ought to give us some degree of comfort.

To better appreciate the levels of safety accorded to nuclear power plants it might be helpful to provide a safety comparison with machines that are more familiar; namely, cars and airplanes.

At the first level of safety, attention is given to the design itself. Careful quality assurance is important for safety but is also important for the manufacturer who wants to stay in business. All three products (cars, airplanes, and nuclear reactors) provide instrumentation to alert the operator to abnormal conditions, and there is clearly some mechanism installed to stop the vehicle or shut the plant down.

A second level of safety is focused on accident conditions. Most of us have come to accept seat belts in cars (it's even mandated by law in many states), and air bags are beginning to become popular. All airliners have oxygen masks that can be used by passengers if sudden depressurization should occur. Likewise, flotation devices are required for passengers and crew whenever flight paths cross large bodies of water. Nuclear power plants require a redundant (second) power supply (normally supplied by diesel generators) to provide emergency power to safety systems in the event of power failure. Likewise, high-pressure pumps are included as part of an emergency core cooling system in the event the normal pumps fail to work properly. In addition to the above safety features, both the airline and nuclear power industries require rigorous training for their operators. Alas, the same cannot be said for everyone who operates an automobile.

At this point, no additional safety features are offered for cars or airplanes. In a nuclear power plant, however, a third level of safety measures is provided. It is postulated that the largest coolant pipe in the reactor system suddenly ruptures (i.e., total separation of a stainless steel pipe typically two to three feet in diameter with one inch thick walls). Should such an accident ever happen (it never has), the hot cooling water would suddenly depressurize and flash into steam. Systems designed to condense the steam (such as large water reservoirs or huge containers of ice) would then come into action. Finally, the containment system itself provides an additional barrier, to ensure no release of radioactive contamination to the outside air.

After Chernobyl? You've Got To Be Kidding!

87

My thesis professor at M.I.T., Dr. Norman Rasmussen, became an international scientific celebrity in the early 1970s when he headed up the largest governmental task group ever assembled to calculate the risks associated with nuclear reactor accidents. His team was the first to systematically apply a new tool called "Probabilistic Risk Assessment" to the nuclear business. The essence of this method is to determine the probability of an event happening and then multiply this probability by the consequences of the event. The product is the overall risk. While little credence is normally given to the actual numbers attained, principally because the input data for low probability events are hard to obtain, it has become a very powerful tool to compare relative risks.

After performing this study, Professor Rasmussen told me that he now steadfastly refuses to answer the question, "How many people can be killed in the worst nuclear reactor accident?" It's not that he is hiding anything; rather, it's because there are so many variables that go into predicting the consequences of such low probability accidents that an absolute answer is meaningless. When insistently confronted by such a request, the conversation normally goes something like this:

Professor: "To answer your question, I have to get some information from you. How many people can get killed in the worst airplane accident?"

Questioner: "Well, I suppose the worst would be if everyone went down in one of those big jumbo-liners. I'd say perhaps 300."

Professor: "Then the number of deaths in the worst nuclear power plant accident is two."

Questioner: "Two? How can it be only two? I've heard numbers like hundreds, or even thousands. How can you say only two?"

Professor: "You didn't give me the worst airplane accident."

Questioner: "What? I postulated that the biggest commercial airliner I know about went down with a full load. What could be worse than that?"

Professor: "Well, what if two of those jumbo-liners crashed head-on? In that case, my answer is four."

Questioner: "Four? That still seems absurdly low. That's much lower than I've read about in the newspapers."

Professor: "Yes, but you've still not given me the worst airplane accident."

Questioner: "What do you mean? I've assumed that the two largest commercial airplanes in existence have crashed, fully loaded. What more do you want?"

Professor: "How about the possibility of that head-on crash taking place over Dodger Stadium in the middle of a World Series game?"

Questioner: "Well that's ridiculous!"

Professor: "Now you're beginning to get the point!"

After Chernobyl? You've Got To Be Kidding!

89

Professor Rasmussen is right. The professionals within the nuclear power industry normally assume extremely unlikely sets of circumstances in estimating the consequences of worst case scenarios. In fact, compared with the safety standards of any other industry, it does indeed seem to border on the ridiculous. It is only within the context of a conversation like the above that one can put such hypothetical accidents into proper perspective.

THREE MILE ISLAND

During the early 1970s, the American people appeared to be buying into nuclear energy. Utilities were enjoying favorable results from their early plants, and the orders for ever larger units came rolling in. However, by the mid-1970s, increasing restrictions and delays in the regulatory process began driving up costs, and overall liabilities began to mount to the point that orders for new nuclear units came to a halt in 1978.

Then on March 28, 1979, shortly after 4 a.m., a series of equipment failures and human errors resulted in an eventual partial core meltdown of the Three Mile Island (TMI) Unit 2 reactor. Two-inch headlines blackened the front pages of newspapers around the country.

Reporters flocked to Harrisburg, Pennsylvania, a few miles from the plant, in what was clearly the media event of the decade. To be sure, there were groups within the nuclear community itself that didn't help calm things down. The NRC was one of the first to speculate on the possibility of a hydrogen explosion, which might rip open the containment structure. Needless to say, that speculation certainly added to the media frenzy. I was not spared the concerns expressed so articulately by the constant media barrage, and I was over 2000 miles away from the scene!

About three months after the accident, I was invited to attend a World Council of Churches conference, held at

M.I.T. (Cambridge, Massachusetts), titled "Faith, Science, and the Future." The delegates, coming from 56 countries, consisted of scientists, theologians, and social scientists.

During the flight from Seattle to Boston, I scanned some of the reading material provided by the conference organizers and noticed a scathing article written by Rev. William Mielke, a pastor from Harrisburg. I had heard of Rev. Mielke because he was the person hand-picked by William Thornburgh, then governor of Pennsylvania, to lead the children and pregnant women of the Harrisburg area to emergency shelters, as a precautionary measure after the accident at Three Mile Island.

Perhaps more than any other person, Rev. Mielke lived the emotions of that accident. He wrote about some of the irrational behavior in which he found himself immersed, as a direct result of the trauma. For example, he reported that one of his fellow clergymen (with whom he worked at the Red Cross shelter) admitted that during the evacuation he had put an extra gas can in the trunk of his car, along with a hammer, screwdriver and axe. No particular logic, simply the survival instinct. I thought it best to avoid Rev. Mielke at the conference if at all possible.

Midway through the conference, a sociologist gave a plenary address in which he concluded that the human species was going to die for lack of energy supplies. It was the most depressing speech I have ever heard. During the question and answer period I was able to get the floor long enough to ask the speaker if I had missed something: "I didn't hear you discuss the standard fossil fuels, nor, in fact, did I hear you even mention solar or nuclear. Did I miss something?" "No," the speaker responded, "you didn't miss anything. I've considered all of them, and nothing makes sense. Our energy pool is about to run dry."

I was stunned. The session was declared over, and as I struggled to my feet I was approached by a gentleman who looked like he was about to take me to task. I

After Chernobyl? You've Got To Be Kidding!

91

glanced at his name tag and almost collapsed—Rev. William Mielke!

There I was, having obviously identified myself as "pro-nuke" by having the gall to ask the closing speaker if he had considered nuclear energy. Trying to regain my composure, I was confronted with the person who I was certain must be the most livid anti-nuclear member of the human race. I wondered how long it would take before my carcass would be hanging from the nearest rafter.

To my complete astonishment, the pastor was applauding my courage for bringing up the topic, particularly so close to the aftermath of the "disaster" itself. If there was any attack at all, it was an admonishment that those on the inside of the nuclear industry should be doing more to tell the positive side of the story. I couldn't believe my ears. Why would a person like Rev. Mielke, who had clearly walked through the darkest valley brought on by the nuclear industry in this country, be so positive about nuclear energy?

The reason is that because of his instant notoriety, Bill Mielke could get into any corporate office or to any policy making group in the land, including the top echelons of government. He started asking some hard questions, such as: What actually happened and what were the real health effects? Where are we going to get replacement electricity? What are the replacement power costs? What are the risks associated with that replacement power?

The answers he got convinced him that the best thing to do was to build more nuclear power plants. Any other option would cost more, both in dollars and health risks to the public.

Well, what actually did happen at Three Mile Island, and what were the actual health hazards to the public?

The only significant discharge of radioactive material was Xenon-133, and a trace amount of Iodine-131 (both gases).

This resulted from some radioactive steam produced by contaminated cooling water that escaped the containment building due to a ruptured overflow system. The average radiation dose to the inhabitants of Harrisburg was approximately 2 mrem, which represents well under 1% of their annual natural background radiation dose.

The maximum dose that a person could have received, had he or she stood for a full month flat against the containment building, would have been about 80 mrem (about 20% of the normal yearly dose).

It is clear that nobody was killed or even injured during this accident, labeled as the worst nuclear power plant accident in U.S. history. This is quite a different result than one might have expected from the horror stories that filled the front pages of our leading newspapers for well over a month. In fact, many of the reporters who flew into Harrisburg actually received more radiation (from additional cosmic rays at their flight altitudes) than the residents of the area surrounding the plant.

Despite the low levels of radiation released as a result of this accident, there is an infinitesimal chance that the cancer rate could increase slightly in the area. Using the linear hypothesis theory outlined in chapter 4, health physicists calculate that somewhere in the span of the next half-century, there could be possibly one premature death due to cancer within a 50-mile radius of the TMI plant. This calculation, we recall, is based on the theory that any level of radiation, no matter how small, has an injurious effect. Consequently, this projection almost certainly represents an upper limit. To put this number in perspective, there will be over 30,000 premature deaths in this same 50 mile radius over the next 50 years from non-radiation induced cancer. How many times did that number appear in the media?

Yet the fears generated by this accident are very real. Several months after the accident, I received a call from a

After Chernobyl? You've Got To Be Kidding!

93

former high school classmate. She said her daughter was planning to fly from the west coast to New York and her plane would take her within a few hundred miles of Three Mile Island. Would she be safe? Hopefully, a recap of the actual hazards noted above will give you a clue to my answer.

Before leaving this section, I should note that those who refer to Three Mile Island as a "disaster" are partly right. It certainly was not a disaster from the standpoint of adverse health effects to the public, but it was very much a disaster from an economic point of view. General Public Utilities, owner of the plant, spent over one billion dollars just to clean up the mess, to say nothing about power revenue losses. Consequently, the industry was quick to establish a "lessons learned" initiative. Both the Institute of Nuclear Power Operations (INPO) and the Nuclear Safety Analysis Center (NSAC) were created by the utility industry. The result was the institution of new communication paths to alert all reactor owners to possible problem areas and to improve operations at all levels. Based on key performance indicators, the industry is now in a much stronger position to prevent the kind of accident that occurred at TMI.

Yes, TMI represented a huge financial disaster, one that the industry is making enormous efforts to ensure will not happen again. However, from a public health point of view, any truly objective person would have to conclude that TMI stands as a monument to the safety of nuclear energy.

CHERNOBYL

The accident at the Ukrainian Chernobyl reactor was quite a different matter. On April 26, 1986, two detonations (steam explosions) rocked reactor Unit 4 at Chernobyl. Flames of burning graphite shot into the air, and a radioactive cloud billowed into Western Europe, followed by detectable fallout literally around the globe.

This was, without doubt, the worst nuclear power plant accident in history. The Chernobyl accident was truly the "what if" scenario that reactor safety engineers have worried about for the past half-century. We call it an "end of spectrum" event because it combined almost the worst set of circumstances conceivable and had no containment.

As I'm sure most readers are by now aware, many of the former Soviet reactors do not have containment systems, in stark contrast to power reactors in all countries of the Western world. Their attitude, prior to this accident, was that containment was unnecessary.[3]

We now know that 31 people were killed outright as a result of this accident. (Twenty-nine were firefighters.) Although most of the immediate deaths were due to burns from the fire, the radiation doses received by these workers were high enough that they probably would have died from radiation sickness. In order to minimize the latter possibility, gallant efforts were made by medical teams, including the U.S. bone cancer specialist Dr. Robert P. Gale of the UCLA School of Medicine, to perform several bone marrow transplants. Unfortunately, the success rate was low, principally because of the overriding damage done from second and third-degree fire burns.

Beyond these initial deaths, the two biggest concerns remaining in the aftermath of Chernobyl are the land contamination and the possibility of cancer deaths.

The land contamination is clearly very real. More than 100,000 residents within an 18-mile radius of Pripyat, the town adjacent to the disabled reactor, have been evacuated, and only a few (about a thousand original residents) have returned. Intense decontamination efforts have been underway for several years, but the numerous scientific teams that have toured the site since the accident have not been told by the state authorities if and when large-scale resettlement may be permitted.

After Chernobyl? You've Got To Be Kidding!

95

By way of perspective, however, the actual appearance of the terrain surrounding the Chernobyl site is far from bleak. My wife and I toured the Chernobyl nuclear power station in May 1995 and were pleasantly surprised at the abundance of trees and foliage. While not denying the detectable surface radiation still existing in many areas, it was clear to us that fears of "perpetual uninhabitable cont-amination" are simply not substantiated.

The possibility of an increase in cancer is not clear. Shortly after the accident, a meeting of experts sponsored by the World Health Organization speculated that perhaps 1600 additional deaths by radiation-induced cancers might occur in Europe over the next three decades. Since that time I've heard numbers ten times larger. Such projections are based on the linear hypothesis model outlined in chapter 4, which most health physicists believe is very conservative (i.e., the actual number of deaths would be far lower). Furthermore, in the population and time sample consid-ered by the World Health Organization above, there would be more than 120 million cancer deaths expected from other causes (i.e., non-radiation related). In other words, Chernobyl might be responsible for about one additional cancer in every 75,000 cancer cases.

The highest incidence of cancer would be expected to occur near the site, where the radiation intensity was the highest. An additional factor is that the local population of Pripyat (approximately 45,000 residents) was not evacuat-ed until 36 hours after the accident. In fact, evacuation of the remaining population within an 18-mile radius of Pripyat was not completed until 10 days later.

Dr. William Bair, an internationally renowned health physi-cist, was one of the few Americans to attend the August 1986 meeting in Vienna where the Soviets presented their first scientific disclosures of the accident. Dr. Bair told me after the meeting that the death rate in the Soviet Union due to cancer was 13% at that time (i.e., 13% of the Soviet population dies from cancer, as opposed to approximately 20% in the United States). The delegates were told that the 13% figure could go to 13.2% in the immediate vicinity of Chernobyl. The change in the country-wide figure would not be perceptible.

Numerous early reports coming out of Kiev, a major population center about 50 miles south of Chernobyl, provided great cause for alarm. Actually, the residents of this city received a radiation dose below the value allowed as the annual dose for U.S. radiation workers (i.e., 5000 mrem). In the European countries with the highest exposures, the radiation dose was no more than residents would have received by spending a few months vacationing in Colorado or the French Alps.

Nevertheless, several scare stories propagated throughout Europe. The Swedish government slaughtered reindeer herds in northern Sweden on the grounds that the reindeer were so contaminated with radioactive Cesium-137 as to be unfit for human consumption. A group of radiologists from a Stockholm hospital criticized the reindeer "contamination" standard as irresponsible and politically motivated. They pointed out that thousands of Swedish homes were plagued by high concentrations of radon gas. Such home-owners would have to eat sixty 12-ounce "contaminated" reindeer steaks every day for a full year to expose themselves to the same radiation dose as they received from the radon gas in their own homes. A health physicist from the Riso National Laboratory in Denmark observed that consuming 154 pounds of reindeer meat contaminated to 500 times the Swedish standard would present a risk equivalent to smoking one cigarette per week. Finally, as the hysteria died down, the Swedish government reevaluated its radioactive cesium-ingestion standard. They raised it to a level that would allow each Swede to consume some 1460 pounds of "contaminated" reindeer meat.[4]

The most systematic study to date concerning the health effects resulting from the Chernobyl accident was concluded

After Chernobyl? You've Got To Be Kidding!

97

in 1991 by the International Atomic Energy Agency (IAEA).[5] At the request of the Soviet government, an international team of about 200 scientific experts, representing 25 countries and 7 multinational organizations, convened under the chairmanship of Dr. I. Shigematsu, head of the Radiation Effects Research Foundation (RERF) in Hiroshima, Japan.[6]

This group did not focus on the 100,000 plus people evacuated from the 18-mile "prohibition zone" around the Chernobyl site. Likewise, it did not include the large number of emergency personnel ("liquidators") who were brought into the region temporarily for accident management and recovery work. Rather, they focused on the region beyond the 18-mile circle, where the Ukrainian and Belorussian people have continued to reside. This was in direct response to the USSR's request for international scientific opinion on safeguarding the health of the public who continued to reside in the surrounding regions.

After over a year of exhaustive study, the group concluded that there were "no health disorders that could be attributed directly to radiation exposure." However, the group did find substantial negative psychological consequences in terms of anxiety and stress because of continuing high levels of uncertainty. Tensions were compounded by socioeconomic and political changes occurring in the former Soviet Union. Contrary to earlier speculations, the data did not indicate a marked increase in the incidence of leukemia or other forms of cancer. While the authors of the study acknowledged the possibility of a statistically detectable increase in the incidence of thyroid tumors in the future, "any increases over the natural incidence of cancer or hereditary effects would be difficult to discern, even with larger and well designed long term epidemiological studies."[7] This conclusion, by the best scientific experts in the world, is in marked contrast to what we continue to hear from the mass media.

While visiting Russia in July 1993, I attended a scientific conference in Nizhni Novgorod. The conference, titled "Nuclear Energy and Human Safety," featured several papers on the health effects of the Chernobyl accident. My major impression, after sifting through all the controversy, was that the actual health consequences of Chernobyl may never be known. The majority of the people in the affected area were living in such impoverished conditions prior to the accident that they had never seen a doctor. After the accident, when medical teams arrived to investigate potential adverse health effects, numerous maladies were reported that may or may not have been related to the accident. There was no medical reference point. As a further complication, the economic conditions in the former Soviet Union since the time of Chernobyl have degraded to the point that any excuse for securing funded projects is almost irresistible. Pleas for further medical studies continue to find a sympathetic international ear and funding source.

After talking to several colleagues about the situation at the conference, I was struck by the remarks of a Russian physicist who has studied radiation exposure at numerous former Soviet sites (including those most contaminated) for approximately forty years. He said, "You may find my conclusion a bit ironic, but I believe the accident at Chernobyl has actually saved many lives." Upon noting my surprise, he went on to point out that a major benefit of the medical attention in the downwind regions was to bring medical assistance to masses of people that were dying prematurely because of the lack of even the most basic methods of personal hygiene. It is, of course, most unfortunate that such a traumatic event had to occur to provide this basic service, but his comments do help provide some perspective.

At the time of my visit to the Chernobyl site in May 1995, the controversy over health effects from the accident still lingered. Medical results for the some 600,000 "liquidators"

After Chernobyl? You've Got To Be Kidding!

99

(those brought in to bring the accident under control) are still largely speculative, although the leading causes of death from such personnel are known to be unrelated to radiation. No comprehensive study has been done on this population. In fact, it is unlikely that such a study will ever be done since these people are now widely dispersed throughout the former Soviet Union. Yet the medical director at Slavutich, the new city built to house the workers at the Chernobyl nuclear power station, indicated in response to direct questioning that he and his medical team have not identified a single death in the vicinity of Chernobyl (beyond the original 31) that could be attributed to radiation from the accident.

Although the public health consequences of the accident surely have been exaggerated, the fact remains that it was a very serious event. Lives were lost, and there will likely be more premature deaths as a result of cancer for those persons who resided or worked close to the accident.[8] So we have to ask the question, can it occur again?

I would be less than honest if I didn't confess to a worry here. There are still many reactors of the basic Chernobyl design (without containment systems) operating in the former Soviet Union. On the other hand, I do know that the ex-Soviets are paying a lot more attention to safety since the Chernobyl accident. Having interacted with these scientists for nearly two decades, I have noticed a pronounced shift in their attitudes since the 1986 tragedy.

In the early 1980s I co-authored a book on fast breeder reactors. The publishing contract specified that the text was to be translated into Russian. I later learned from a colleague at the Kurchatov Institute in Moscow that a friendly competition ensued as to which organization in the Soviet Union would be granted the "privilege" of doing the translation. During a 1990 visit to Obninsk, the center of Russian fast reactor research activities, I was told that the number of copies of my book distributed within the Soviet Union is larger than all the English versions sold in the Western world. Whether true or not, I was flattered to learn that I am fairly well known in that part of the world.

Consequently, I should not have been surprised (though I was) to have been approached by a Soviet delegation only a month and a half following the Chernobyl accident during a conference on the island of Guernsey. They were interested in establishing a much closer working relationship with reactor safety scientists in the United States. While I can't say that this change in attitude will produce instant results, operating procedures and safety devices have already been improved. They are now discussing safety precautions in new designs that even we in the West have thought to be unnecessary. The change is dramatic.

The Chernobyl reactor simply could not have been licensed in the United States, or any other Western nation. The differences between the Chernobyl reactor and those designed in the West are many, but there are two fundamental differences that need to be clearly understood.

The first difference relates to basic physics. The Chernobyl reactor has what is known as a "positive coolant coefficient." This means that as the temperature of the coolant rises, the nuclear reaction rate (and, therefore, the power level) also rises. This is clearly undesirable. It would be like designing a car engine that would automatically speed up in direct response to a loss of coolant caused by a rupture in the radiator hose. During the Chernobyl accident, an initial mishap caused the coolant temperature to rise. That in turn caused the power level to increase, further causing uncontrolled steam pressures to blow the reactor apart.

This type of reactor design is not permitted in the West. Great care is taken in the design process to ensure that a rapid thermal change, caused by any number of possible initiators, will not subject the reactor to a runaway mode. The response of the reactor to any accident condition must be to move toward automatic shutdown.

After Chernobyl? You've Got To Be Kidding!

101

The second fundamental difference is the containment system. As stated earlier, until recently the Soviets had steadfastly refused to include containment systems in their reactor power plants, while the multiple barrier philosophy in the West has always dictated the use of such systems. There has never been a reactor commercially licensed in the United States that has not had a containment system. The radiation dose to the public would have been much greater at Three Mile Island if that reactor had lacked containment. Fortunately it had one, and the system worked.

NEW REACTOR DESIGNS

We noted earlier that there has not been an order for a new nuclear power plant in the United States since a year before the Three Mile Island accident, over 15 years ago. Can we now build plants that are even safer?

The answer is yes. From the experience gained to date, and the genuinely remarkable safety record that has been compiled in our nation, many members of the nuclear industry would say that there is no reason to incorporate more safety into the designs. They would claim that nuclear power plants are already much safer than competing electric power plants. Perhaps so, if truly compared objectively. However, my experience suggests that in order for nuclear energy to make a comeback, the new plants need to do even better than their predecessors in order to restore public confidence.

Fortunately, the bulk of the industry agrees with this point of view, and substantial progress has been made over the past decade. Perhaps the biggest step has been to take advantage of "passive," as opposed to "active" features. By this I mean developing a design that works with the laws of nature, such as gravity. These processes will always be available no matter what else might fail. This is in contrast to relying too heavily on active systems, which require power or moving parts to accomplish the mission. The result is a simpler, safer design.

Three very different designs have emerged, all of which take advantage of passive features. The first is the familiar water-cooled reactor. This is the type of plant that currently produces about 21% of the electricity we now use in this country. There are over 100 of these plants currently in operation in the United States. The new idea is to design smaller sized plants (producing only about half of the power of the largest plants currently running) and to place a huge reservoir of water above the reactor core. In the event of insufficient cooling, water would automatically flow from this reservoir into the reactor under the force of gravity. No pumps, which require power and elaborate valving systems, would be necessary. The reactors could be cooled for at least three days with this system, giving operators ample time to arrange for additional cooling water or other safety measures, if needed. As long as the reactor remains cooled, there is essentially no chance of the release of any sizeable quantities of radioactivity.

The second reactor design resulting from the new emphasis on passive safety is the modular high temperature gas reactor (MHTGR). This reactor is cooled by helium gas, rather than water. The fuel is encased in small carbon spheres and then sealed into a graphite block. The advantage of this system is that even if all coolant is lost, there is sufficient heat capacity inherently contained in the graphite mass to absorb the nuclear heat and prevent the fuel from melting and releasing the entrapped radioactive materials. This system should be able to remain cooled essentially forever in the event of an accident.

The third reactor design resulting from the new safety emphasis is the liquid metal cooled reactor (LMR). Liquid sodium is used as the coolant, instead of water. This reactor is quite robust because sodium remains a liquid at low pressure up to very high temperatures (in contrast to water). Consequently, during any thermal upset condition, there is ample time for reactor structural material to heat up and physically expand the core, resulting in automatic

After Chernobyl? You've Got To Be Kidding!

103

reactor shutdown, prior to reaching sodium boiling conditions. This system is capable of remaining cooled for a very long time (weeks, or even months).

Other systems have also been proposed, but are not yet proven to be practical devices. Substantial work has been done on MHTGRs, and LMRs have been built and operated for several decades. In fact, experiments have been conducted using liquid metal-cooled reactors that dramatically demonstrate their ability to shut themselves down during major accident conditions. Although no water-cooled reactors have yet been constructed that incorporate the new gravity-fed passive safety feature outlined above, the technology required is only a small extrapolation from the wealth of experience gained from four decades of operating water-cooled reactors. Furthermore, successful testing of key parts of these new systems has already been completed. Consequently, there is little question that the technique will work effectively.

Fortunately, the strides made in this technology seem to be gaining public acceptance. After publishing an article entitled "Cast Your Vote on Next Generation Nuclear Power" in the April 1990 issue of *Popular Science*, the magazine reported the poll results in their August 1990 issue. Their readers gave a strong vote of confidence to a new generation of nuclear power plants. The results were 80% favorable. This surely cannot be interpreted as a national mandate because it is based on a very small sample of people interested in science. However, I think it suggests that the public would like to see as many safety features as possible built into the next generation nuclear power plants. It is now quite clear that this can be accomplished.

SAFETY IN PERSPECTIVE

Tragic as the Chernobyl accident was, we need to be reminded that nuclear energy is not alone in having inflicted fatal injuries during industrial mishaps.

Perhaps the most publicized large scale accident outside the nuclear industry is the chemical accident that occurred in Bhopal, India, in 1984. That accident resulted in the deaths of over 2,000 people—and severe injuries to more than 200,000 people. The long-term effects are still not known.

Figure 20, constructed from a very thoughtful article published by Robert W. Lee in the March 27, 1989, issue of *The New American*, provides a sobering rundown on the results of accidents in the fossil fuel industry in the 8 years following the accident at Three Mile Island.[4] A list of this type is not to suggest that we ought to cease burning fossil fuels because they represent an undue risk. Rather, it is to suggest that there is no industry in the energy sector that is risk free.

By direct comparison, the nuclear energy option looks remarkably good. Even Dr. Robert Gale, the UCLA bone marrow specialist who dealt first-hand with the most morbid results of the Chernobyl accident, believes that it is time we get serious about our future energy strategy. In a special article in the May 5, 1989, edition of the *Los Angeles Times*, Dr. Gale acknowledges that we must be willing to bear some of the costs of environmental safety regulations within the nuclear energy industry. "However," he writes, "regulations motivated by politics and emotion, rather than science, are an unnecessary expense." He goes on to plead, "If we stop using nuclear energy, we must accept the risks of oil exploration or changes in the environment of the planet that may result from burning fossil fuels. We must understand that simply voting against every energy proposal will not result in a sensible energy strategy."

After Chernobyl? You've Got To Be Kidding!

105

	NUMBER OF DEATHS	EVENT	PLACE	DATE
OIL	70	Underground oil rig explosion	Bohai Bay, China	Jul 7, 1980
	more than 145	Oil tank fire	Caracas, Venezuela	Dec 19, 1982
	more than 508	Gasoline pipeline fire	Cubatao, Brazil	Feb 25, 1984
	40	Bus collision with oil truck	Damagum, Nigeria	Sep 1, 1984
	almost 1600	Passenger ship and oil tanker collision	Mindoro Island, Philippines	Dec 20, 1987
NATURAL GAS	more than 490	Liquified natural gas site explosion	Mexico City, Mexico	Nov 19, 1984
	more than 100	Gas explosion in apartment building	Tbilisi, Former Soviet Union	Dec 2, 1984
	65	Pit explosion	Zaluzi, Czechoslovakia	Sep 3, 1982
	44	Series of mine explosions	Yubari, Japan	Oct 1982
COAL	98	Gas explosion and mine cave-in	Eregli, Turkey	Mar 7, 1983
	36	Explosion and rock falls	Oroszlany, Hungary	Jun 22, 1983
	64	Gas explosion in mine	Natal Province, South Africa	Sep 12, 1983
	62	Mine Explosion	Yubari, Japan	May 17, 1986
	53	Mine Explosion	Guangdong Province, China	Jul 12, 1986
	more than 30	Methane gas fire	Donetsk, Former Soviet Union	Dec 24, 1986

Figure 20. A Sample of Fossil Fuel Accidents in the Eight Years Following Three Mile Island[4]

Nuclear energy is not fail-safe, but if we are willing to strip away the emotion and make an honest assessment, I believe we can reasonably conclude that commercial nuclear energy is not nearly as dangerous as is generally believed. Fortunately, we now have the means to make it even safer.

6

NOT IN <u>MY</u> BACKYARD (NIMBY)

The Nuclear Waste Dilemma

A Skeptic's View

Look. The bottom line with nuclear energy is that it generates waste that lasts for thousands of years or more. I've heard people say it may last for 10,000 years—maybe even a million! Nobody can convince me that we can guarantee a safe way to store such intensely radioactive waste for that long. With the unstable world we live in, anyone arguing for institutional stability for this long is bordering on lunacy.

Furthermore, I don't feel we have any right to pass on this kind of a legacy to unborn generations. We ought to be able to solve the energy problems of today without subjecting our great grandchildren, and their great grandchildren, to the waste problems caused by today's gluttony and insensitivity.

Finally, even if I could be convinced that this radioactive waste could be safely removed from the environment, there is the question of transporting such wastes from the reactors to the processing plants and then to the waste dumps. Hauling that intensely lethal garbage over our nation's public highways and through major population centers is intolerable.

Dealing with the waste is the number one problem of nuclear energy. Even the scientists don't have a solution!

Although there are several aspects of nuclear energy that stir the emotions of many, the nuclear waste issue seems to be the one that most bothers people. Radioactive waste that can last for thousands of years would, indeed, seem to be a monumental problem.

As a prelude to a discussion of the nuclear waste dilemma, I am reminded of a remark Bertram Wolfe once made that has been useful to me in shaping a constructive perspective. Dr. Wolfe, one of the most respected international statesmen in the field of nuclear energy (retired corporate executive with General Electric and former President of the American Nuclear Society), quipped that judging nuclear energy from the standpoint of the waste issue was a bit like judging the merits of parenting from the vantage point of the diaper. Any system (human or mechanical) that results in some useful output also generates waste. So the question is not whether waste exists; rather, it is the net cost of dealing with the waste, both in terms of health effects and environmental impact.

Surprising as it may seem, Dr. Wolfe went on to point out that one of the great advantages of nuclear energy is that the waste problem is so easily solvable. In fact, nuclear energy may be the first large industry in history that is capable of removing essentially all its waste from the biosphere.

How can this statement, so completely contrary to the skeptic's concerns expressed above, possibly be justified?

In order to understand this we need to identify what we mean by the term "nuclear waste," look at the volume and longevity of such waste, and then discuss how to dispose of it. We will then be in a position to assess the waste disposal health risks and go on to consider the waste transportation issues.

WHAT IS NUCLEAR WASTE?

Nuclear waste is generally categorized as high-level and low-level waste. Low-level waste (LLW) derives its name from the fact that its radioactivity is very small. Typical items in this category are filters that have been used to remove radioactive debris from air or water, plus other cleaning utensils such as mops and rags. Gloves, laboratory equipment, instruments, or other components used in a radiation area are also considered low-level waste. The activity, or radioactive concentration, of such low-level waste is about one billionth of that of high-level waste. However, the volume of such waste is thousands of times larger.

Approximately 60% of all low-level waste is a by-product of nuclear power plant operation.[1] The remaining 40% comes from a wide variety of the some 20,000 civilian establishments that have licenses for dealing with radioactive material, principally hospitals, research and industrial institutions, and universities.

In the past, low-level waste was transported to approved sites and buried in shallow trenches without a great deal of special expense or concern. Because of the attention that has been drawn to a few isolated mishaps (essentially all of which have resulted in only infinitesimal and non-harmful releases of radiation), public pressure has imposed much stricter regulations. The number of licensed low-level repositories in the U.S. is now very small (currently only three), and this has caused a good deal of consternation—particularly within the medical industry, which is rapidly increasing its use of radionuclides for diagnostic and therapeutic purposes.

High-level waste (HLW) is normally defined as the nonusable radioactive waste generated during the fission process in nuclear power reactors.[2] The bulk of the radioactivity comes from the fission products, which are produced as the fuel undergoes nuclear fission. As we noted in chapter 4, the unstable fission products decay to other isotopes or elements. The time for such decay varies widely, from a small fraction of a second to several billion years. Since the half-life of most fission products is relatively short, the level of radioactivity generally goes down quite rapidly. After one year, the radioactivity of spent fuel (used fuel assemblies removed from a reactor) is only 1% of its original activity.

Strontium-90 and Cesium-137 are produced abundantly during the fission process and have half-lives of 29 years and 30 years, respectively. These two substances provide the highest source of radiation intensity in the high-level waste stream for the first two hundred years or so. After 600 years, the radioactive intensity of these two substances is reduced to only one-millionth of its original value.

The other major component of high-level waste is made up of "transuranics" (TRU). Elements in this category are also generally produced in a nuclear reactor. These elements, principally neptunium, plutonium, americium, and trace amounts of even heavier elements, typically have long half-lives. Consequently, even though they provide a low source of radioactivity, these elements (like radium) can last several thousand years.

Since it is the high-level waste (HLW) that must be isolated from the biosphere for several centuries, the remainder of this chapter will be focused primarily on this material.

HOW MUCH NUCLEAR WASTE IS THERE?

Without question, the principal redeeming feature of nuclear waste is that the volume of HLW is very small. This is not to suggest that HLW poses no problem, because, indeed, it does. However, it is important to recognize that the quantities we need to deal with are quite tractable, much smaller than the waste of any comparable industrial endeavor.

A large modern nuclear power plant, which generates enough electricity to supply a city of a million people, produces only about one ton of HLW per year. The volume of this waste material is approximately two cubic yards (less than the space occupied by six file cabinets or the back of a pickup truck). If Americans received all their electricity from nuclear energy, rather than the 21% we receive today, the amount of HLW we would each be responsible for on an annual basis could be contained in three small marbles. With waste quantities this small, logic would tell us that there must be a way to achieve proper disposal.

In comparison, a like-sized coal plant (our principal source of electricity today) generates waste that, by weight, is 5 million times larger than nuclear. This waste consists mainly of fly ash and the gaseous products carbon dioxide, sulfur dioxide, and nitrogen dioxide. It's billions of times larger in volume than nuclear waste. Further, because of the trace amounts of uranium and radon in all coal deposits, the fly ash from a coal plant discharges appreciably more radioactivity into the stack and into the surrounding environment than the radiation released from a comparably sized nuclear plant. Finally, a thick slurry that remains as a residue from the scrubbers, installed to prevent much of the sulfur dioxide from going up the smokestacks, constitutes a waste volume that is 10,000 times larger than the HLW from a sister nuclear plant.

To put the above comparisons into perspective, the United States now produces over 36 million tons of hazardous chemical waste per year. If we assume that the population is 250 million, that means each of us is responsible for nearly 300 pounds of hazardous waste per year, all of which needs careful disposal. The three marbles of HLW from the nuclear industry represent a remarkably small increment. Over our average lifetime of 70 years, we each accumulate over 11 tons of hazardous waste. Assuming a heavy dependence upon nuclear energy, our HLW lifetime accumulation would, by comparison, fit within a frozen fruit juice can 5 inches high by 2-1/2 inches in diameter.[3]

By any relative measure, the volume of HLW that we must deal with is small, incredibly small.

Hazardous Waste Accumulated during One's Seventy-Year Lifetime

HOW LONG DOES IT LAST?

Yes, you may be saying at this point, but to imply that small volumes solve the problem is misleading, because high-level nuclear waste lasts for a long, long time!

Indeed, some of these wastes do last for a long time. In contrast to the other hazardous substances, however, they don't last forever. In time they totally decay away. Arsenic, chlorine, mercury, and lead, for example, never decay. They remain in the biosphere for not just 1000 years, or 100,000 years, but for eternity. We insist that nuclear waste be isolated for 10,000 years, yet no such restrictions are placed on toxic industrial waste sites where containers could (and sometimes do) fail in only a few years.

If we reprocess spent fuel and extract the unused uranium and transuranics for subsequent use, the radiotoxicity of the remaining HLW falls below that of coal fly ash after about 400 years. In just over 1000 years, the HLW is less hazardous than the uranium ore from which it was originally extracted.

Schmidt and Bodansky, in their thoughtful book, *The Energy Controversy: The Fight Over Nuclear Power,* put it this way. Let's suppose we received all our electricity from nuclear energy for the next two decades (five times the amount we currently receive). After 600 years of burial, the HLW from these plants (residing some 2000 feet beneath the surface of the earth) would produce less than 0.1% (.001) of the total radioactivity naturally residing in the top 2000 feet of the earth in the United States. If our concern is radiation, this would say the risks from natural radioactivity in the ground would be at least 1000 times larger than that from the HLW deposited there from a mammoth nuclear power industry. And this comparison is based on the assumption that the HLW would be evenly distributed in the ground, rather than in carefully sealed canisters embedded in geologic media known to be stable for millions of years. After 1000 years, the total increase in natural radioactivity in the above example would be one part in a million.[4]

We see, then, that the popular belief that we have to ensure absolute insulation of this high-level waste from our biosphere for thousands of years is erroneous. First, there is very little of it to begin with, and second, after a thousand years or so, most of what we started with has literally disappeared. It no longer exists. This is very much in contrast to other hazardous waste.

HOW IS WASTE DISPOSED OF?

The nuclear industry has long recognized that some method of high-level waste disposal would be needed eventually. This topic was not given high priority in the early days of nuclear energy because there was never a serious technical concern about being able to design and implement a safe, long-term storage system for such small volumes of material.

In retrospect, this casualness on the part of the industry was a serious political mistake. Scientists were not concerned because they believed the problem was straightforward, and didn't require intensive research and development early on. Unfortunately, the delays have sent the opposite message to the public; namely, that there isn't a solution.

The disagreement in the technical community today is not whether there are ways to safely dispose of HLW. There are. It's an argument over which is the best way. Given the numerous possibilities for handling a waste form of such small volume, it should not be surprising that creative scientists and engineers can come up with a variety of possibilities, all of which would provide levels of safety for the public far exceeding what should be reasonably required.

No matter which method of waste disposal is proposed, the bottom line requirement is to be sure the radioactive waste cannot reach and contaminate our drinking water. All other ways for the nuclear waste to reach and injure the human body are small in comparison to that of contaminated drinking water.

The first step in disposing of HLW is to take the spent fuel elements out of the reactor and store them in cooling basins (confined pools of water) for about three years.

This step allows the radioactivity to fall to a small fraction of 1% of the level existing inside the reactor. These cooling basins typically contain 40 feet of water, surrounded by six-foot-thick reinforced concrete walls lined with an inner steel jacket. Although these storage basins are very safe and provide excellent environmental protection, such storage space is limited in size. They were never intended for permanent storage.

The only real controversy within the scientific community occurs at the next step. Most professionals prefer to recycle (reprocess) the spent fuel elements in order to extract the full energy content of the fuel. A substantial amount of usable fuel remains in the spent fuel elements (principally uranium and plutonium). It can be chemically extracted in a reprocessing plant and recycled into new fuel elements. This is done in a manner analogous to recycling used aluminum cans into new aluminum products. In addition to recycling used fuel, useful products for medical and industrial use can also be obtained. During the same reprocessing step, the undesirable fission products are collected and put into a very stable, low volume, solid waste form (usually glass) for long-term storage. This reprocessing approach is now in commercial operation in France and Britain and soon will be in Japan.

In contrast, some professionals argue that the cost of providing such recycling exceeds the value of the fuel recovered. Although this is solely an economic argument (which can change with time), additional concerns over the appropriateness of using plutonium for fuel (see chapter 3) have led the United States to adopt a "once-through" policy. We have elected to send the spent fuel straight to a repository. But because no repository for storing HLW currently exists, utilities are being forced to build more on-site storage facilities to retain their spent fuel. Since the volume of these spent fuel assemblies is relatively small, this practice can likely be tolerated for several more years. However, at some point the industry will become severely "constipated," and a repository (or some other avenue for removing this burden from the utilities) will be necessary.

Assuming selection of the recycling path, a vitrification plant is normally built to turn the HLW into glass. Glass has been almost universally chosen as the preferred waste medium because it is known to be very stable (i.e., it dissolves exceedingly slowly in water). Samples of glass from the Babylonian era (3000 years ago) have been found fully submerged in water and yet still intact. These are glass samples made with far less technical sophistication than now possible.[5] There are studies from Canada that suggest it may be possible to make HLW glass forms that could last 100 million years. A conservative estimate for dissolution of glass into invading water is at least ten thousand years.

The next step is to provide a canister into which the glass is placed. The canister is made of highly corrosion resistant metal, specifically designed to maintain its structural integrity for very long periods of time in a damp atmosphere. A principal design requirement for the canister is to prevent water penetration for at least a thousand years.

Figure 21 provides an illustration of how the waste package might fit into a permanent storage location. An enclosing material (overpack), typically fabricated from steel or titanium, surrounds the canister to provide a further environmental barrier. This, in turn, is surrounded by crushed rock to fill in the cavern in the geologic host material. The rock or other backfill material that surrounds the manufactured waste package is carefully selected to ensure very low water solubility, should water ever reach that point.

Certainly the most publicly visible factor in the overall HLW disposal process is the selection of the host repository. The essential features that determine the suitability of a site include seismic stability (to minimize the potential for earthquakes allowing water to invade the repository), hydrology (to make sure it is difficult for water to access or seep away from the waste form), and geologic stability.

Since the principal purpose of the repository system is to prevent the HLW from contaminating drinking water, the suitability of a repository is judged on the basis of how well this protection is supplied.

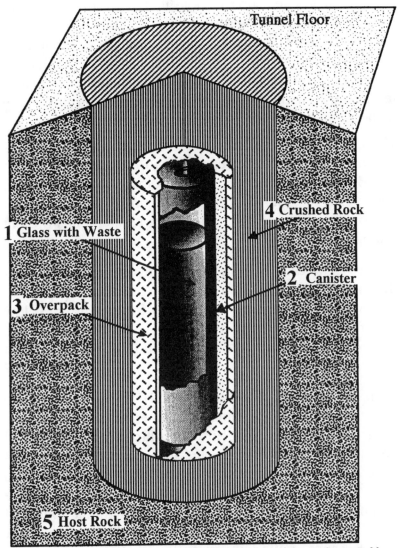

Figure 21. Typical High-Level Waste Package Sealed Deep Within a Stable Geologic Formation[6]

The absence of groundwater is a highly desirable feature of a repository site. Without water the waste does not move. Another attribute is a geological medium that is insoluble. Granite, basalt, and clay are excellent candidates. The only way for water to get to the waste package in a short period of time is massive cracking of the host material. It is interesting to note that salt formations have also been studied for suitability as HLW repositories. Clearly, salt fails the insolubility test, but the fact that there are salt formations that have remained intact for millions of years demonstrates the lack of water. Numerous sites, including a wide variety of geologic formations, have demonstrated such long-term tranquility.

The United States has narrowed its search for the nation's first high-level geologic repository to the Yucca Mountain area in southern Nevada. The geologic medium is tuff, which is a volcanic ash compressed under its own weight and often tightly welded by the high temperatures present during its formation.

Figures 21 and 22 help to illustrate five sequential protective barriers discussed so far: 1) insolubility of the glass waste form, 2) corrosive resistance of the canister, 3) low water solubility of the overpack, 4) insolubility of the crushed rock, and 5) lack of ground water in the host repository media.

Even in the remote event that some HLW does get into the groundwater, such contaminated water would take a very long time to reach the earth's surface. A typical burial depth for the HLW, as depicted in the representative geologic repository illustrated in figure 22, is well over 1,000 feet. At any of the repository sites under consideration, the groundwater flows very slowly, typically less than one foot per day. For the Nevada site, it is only about an inch per day. Because ground water prefers to flow horizontally, rather than vertically, a distance of some 50 miles would be necessary for this water to reach the surface, resulting in a travel time of well over 1000 years.

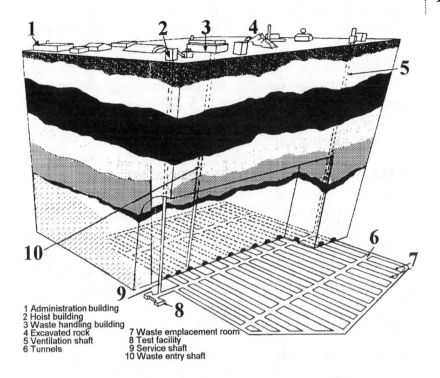

1 Administration building
2 Hoist building
3 Waste handling building
4 Excavated rock
5 Ventilation shaft
6 Tunnels
7 Waste emplacement room
8 Test facility
9 Service shaft
10 Waste entry shaft

Figure 22. Schematic of a High-Level Waste Geologic Repository Layout[6]

Finally, most of the radioactive material in the HLW package would not move with the velocity of ground water. Rather, it would be constantly adsorbed on the rocks and other ground material as the ground water slowly percolated through, so that its average transportation time would be less than a hundredth of that of the flow of water. This effect could add tens of thousands of years to the time when any appreciable amount of radioactive waste matter could get into the drinking water. Furthermore, most of this material would precipitate out of solution and become a permanent part of the rocks deep within the earth.

Any one of these barriers provides a very substantial layer of protection for the ground water in the vicinity of a repository. Taken together, the effect of these individual barriers directly multiplies, such that the probability of any groundwater contamination is exceedingly small. Even if some small amount of the waste material should reach the surface after ten thousand years or so, any radioactive contamination in the drinking water would be so low as to be completely masked by the low level of natural radioactivity that is always present. From our earlier discussions, we know that after such long time periods there is essentially no radioactive waste material still in existence. Almost all of it has long since decayed away.

WHAT ARE THE RISKS OF WASTE DISPOSAL?

So what's the bottom line? Can nuclear waste be safely stored? Yes, it can. The quantities of high-level waste are small, and the waste material itself completely disappears over a long period of time. Consequently, the task of designing and constructing a geologic repository capable of providing monumental layers of protection to the public is really quite straightforward.

But what would happen if, somehow, the contents from the HLW package *did* get directly into our drinking water? Would this be a calamity of untold proportions?

Indeed, direct ingestion of undiluted HLW would be exceedingly dangerous. It may come as a surprise, however, that there are many substances used in daily commerce that are far more dangerous to human life than nuclear waste. Dr. Bernard Cohen pointed out that if Americans consumed all the domestically produced barium, there would be 10 times the number of deaths than if we consumed all our nuclear HLW.[7] On a similar comparative basis, it is even more sobering to note the number of deaths (in comparison with HLW) for consuming the following domestically produced substances: ammonia and hydrogen cyanide (600); phosgene (2000); and chlorine gas (40,000).

Furthermore, the health hazards of HLW diminish with time because of the radioactive decay process. In contrast, the health hazards of most nonradioactive substances remain the same forever. For example, at the time nuclear waste is first placed into a repository, a pound of HLW is about ten times more hazardous to humans than a pound of arsenic trioxide. However, after 100 years the HLW has decayed to the point that the relative hazards are about equal. After 600 years, that same pound of HLW is only one-tenth as hazardous as the pound of arsenic trioxide (whose properties have not changed). In more familiar terms, a pound of copper is as lethal to the human system as a pound of HLW that is 600 years old. That same copper is commercially used to carry our drinking water directly to our faucets. How many times have you read about that in the newspaper?

Putting such factors together, Dr. Cohen calculates that the number of deaths attributable to the waste generated by nuclear energy is 0.018 per year per plant. To put this number in perspective, the comparable number for coal is somewhere between 5 and 25 deaths per year per plant (depending upon the degree of rigor achieved in scrubbing the combustion gases that go up the coal plant stacks). In other words, if our interest is in comparing public health hazards resulting from the waste generated by nuclear energy and coal, nuclear turns out to be at least 1000 times more safe.

For those still skeptical about such numbers, since they are admittedly calculations based partially on repository systems that have never been built, it may be comforting to know that a full scale "experiment" has already been conducted. The "repository" performance from this natural incident can be observed some 1.8 billion years after the high-level nuclear waste was "put in place."

A seam of high grade uranium located in Gabon, West Africa, consisted of sufficient quantities of uranium naturally enriched with U-235 that a fission chain reaction occurred some 1.8 billion years ago.[8] This natural reactor operated for several hundred thousand years, generating about 10 tons of high-level nuclear waste. After thoroughly investigating the site, scientists from French and U.S. laboratories concluded that most of the solid fission products and essentially all the transuranics remained locked within the deposit, and have subsequently decayed to become harmless. We emphasize that this waste was generated under natural conditions within only a few feet of the surface of the Earth and had no engineered barriers to prevent the transport of waste products. Yet there is a wealth of evidence that essentially all of the waste materials remained very close to where they originated.

This direct evidence should help those who may still question the effectiveness of the repository approach in isolating high-level waste from the biosphere.

HOW ABOUT TRANSPORTATION RISKS?

Another problem associated with the nuclear waste issue is transportation. The volume of low-level waste is rather substantial, and a transportation system must exist to move this material to an approved low-level waste storage site. For high-level waste, it is possible in the "once-through" cycle that the only shipments needed are those to transport the spent fuel to the repository. If fuel recycling is involved, more transport steps are necessary.[9]

Recently, I read about citizens in a mid-sized Kansas town that became outraged when they learned that a train carrying nuclear waste came within a quarter mile from a school playground. They simply couldn't believe that any company or government agency could be so insensitive as to expose small children to such imminent danger.

How risky is the nuclear waste transportation business? Shipments of radioactive material have been a normal part of our transportation commerce for nearly a half century. Approximately 100 million radioactive packages have been shipped during this time period without a single radiation-induced injury to any member of the public. For example, from 1971 to 1981, 30 million radioactive packages were handled. Packaging failures occurred in only 58 of these cases (a success rate of 99.9998%), and they were all associated with low-level waste shipments. The health consequences associated with low-level package failure are so minimal that there is no logical incentive to design these packages to withstand severe accidents.

High-level waste packages, on the other hand, have exceptionally stringent design requirements to ensure that waste package integrity is maintained under all conceivable conditions. The high-level waste package (shipping cask) must be shown to survive the following damage tests: a 30-foot fall onto a flat, unyielding surface; a 40-inch drop onto a 6-inch diameter vertical steel rod; exposure to a 1475°F fire for 30 minutes; and submersion under 50 feet of water for 8 hours.

The design must survive these trials, in sequence, with essentially no loss of containment and no significant reduction in radiation shielding.

Other tests of these high-level shipping casks have been made to verify their durability. In one demonstration, a spent fuel cask was mounted on a truck and rammed into a concrete wall at 80 miles per hour. The collision destroyed the vehicle but only slightly damaged the container. In another test, a 120-ton locomotive smashed broadside into a cask at 80 miles per hour. The impact demolished the train but hardly dented the waste cask. Of the hundreds of thousands of high-level waste shipments from 1971–1981, 48 were involved in transportation accidents. Yet not a single package failed, and there was no release of radioactivity.

It is probably worth noting that there are about one hundred million shipments of all types of hazardous materials—flammable, toxic, explosive, corrosive—moving on U.S. roads and tracks every year.[10] This is in addition to the millions of tons of oil, fertilizer, etc., that is shipped through our inland waterways and coastal harbors. We are all too familiar with the environmental contamination caused by oil tankers running afoul. In comparison, the nuclear waste traffic is small, and much better protected.

As a final note of reassurance, high-level waste cask designers have even assessed the issue of sabotage. The results indicate that if saboteurs were to attack a HLW package in the middle of New York City during the height of rush hour traffic using whatever equipment they desired and could transport to the scene, the maximum damage they could do would release 0.0006% of the contents. Translated into public health consequences, this could cause at most one cancer death in the greater New York City area in a 30-year period. For comparison, 250,000 cancer deaths will occur from other causes in this geographic area and time period.

The transportation of nuclear waste cannot be said to be completely risk free. However, based on the exacting design requirements, results from exhaustive tests, and actual experiences to date, I believe we can assure our friends near that schoolyard in the Midwest that their fears are unfounded.

OTHER WASTE DISPOSAL OPTIONS

One would think that the methods outlined above should be more than adequate for dealing with our nation's nuclear waste. By any scientific measure, the dangers posed to the public are incredibly small, particularly in comparison to the hazards of other industries. Yet once the search for the first geologic HLW repository in the United States began in earnest, public protests have risen to almost a religious fervor. "Not in my backyard!" is the battle cry that has echoed from city to city and state to state. One would think the black plague was being proposed.

Quite frankly, this type of public response was not anticipated by the nuclear community. They have been stunned. While I think most nuclear industry personnel believe that the U.S. government should continue its effort to build the first deep geologic HLW repository, there is a small but growing contingency that is becoming persuaded that other options merit exploration. If the general public continues to oppose building the repository, that is a decision the industry will have to accept, even if this blockage is based on reasoning that is seriously flawed. We live in a democracy, and in the end the public will make its choice. Other options will almost surely be more expensive, but the public may be willing to pay the price.

So what are the other options? Several have been proposed. One possibility is to drop HLW canisters into deep canyons in the ocean that would be conducive to self-burial. Because of the corrosiveness of the surrounding salt water, it would be very difficult to maintain canister integrity for more than a few hundred years, but the glass waste package would still be highly stable. As pointed out earlier, glass does not dissolve readily in water, even sea water. After thousands of years in the Mediterranean Sea, glass objects are still in good condition. However, even if the HLW were completely and uniformly dissolved in the ocean water, the amount of radioactivity added to the natural radioactivity of seawater would be undetectable.[11, 12] The principal argument against this proposal is the almost certain outcry that the nuclear industry is trying to pollute the world's oceans.

Another option would be to place HLW canisters in carefully selected glacial ice fields. The canisters are initially fairly warm because of the relatively high radioactivity from the decay of fission products. Hence, the canisters would melt the ice and penetrate to great depths in the ice fields before stopping.

Because the quantity of HLW is so small, some scientists have proposed sending the waste packages into outer space. Given the proper orbital trajectory, there would be a very low likelihood of the debris ever getting back to earth. However, this proposal suffers from the same potential emotional objection as ocean disposal; namely, the accusation that we would be polluting outer space (even though outer space is already permeated by radiation from cosmic rays). An additional concern is a potential mishap on the launch pad, particularly after the Challenger disaster in 1986.

Another possibility is to store waste packages similar to those designed for an underground repository *above* ground in a specially designed, hardened structure. Those favoring this option point out that we already have experience with such large structures. The pyramids have lasted for over 4000 years without a maintenance program. Since future generations may find useful purposes for what some now call waste, the stored materials would be more easily retrievable.[13] Given advances in technology that are to be expected of future generations, an above ground storage system would allow ready improvements, if necessary. Finally, this option may be relatively inexpensive.

There is yet another option that has undergone cycles of favor and disfavor within the technical community. The scientific name is "partition-transmutation," often referred to simply as "P-T." The term "partition" means to separate chemically the nuclear waste into its constituent parts, batches of materials that have similar properties and potential uses. This is similar to how many of us separate our household garbage; namely, materials for recycle (glass, aluminum, newspapers and computer paper), materials for compost (grass cuttings, fruit and vegetable peelings), materials for burning (wood scraps, tree prunings), hazardous materials for special treatment (crankcase oil, old paints, cleaning substances), etc. Once the nuclear waste is chemically partitioned, the low-level waste products could be easily disposed of in low-level burial sites. Many radionuclides could be used in the commercial sector (for medical diagnostic and therapeutic applications, for sewage sterilization to convert a landfill problem into a non-hazardous usable fertilizer, etc.). Those isotopes with long half-lives could be set aside for further treatment.

Dealing with the long half-life nuclides would be of particular interest. If these materials could be permanently destroyed or changed into useful substances, a great burden could be lifted from the challenge of long-term repository storage. This is where "transmutation" comes in. Studies and limited-scale experiments have been conducted that indicate it may be possible to convert all long half-life radionuclides contained in nuclear waste into benign or commercially useful products.

There are two categories of radionuclides that have long half-lives and therefore cause concern for repository integrity beyond 1000 years. One group is the "actinides," which consists of uranium and all the elements heavier than uranium. The second is the "mobile" fission products, principally Technetium-99 and Iodine-129. They are called mobile because, in contrast to all other nuclear waste products, they tend to travel with water (and, therefore, could potentially traverse substantial distances from a failed repository). If both of these groups are permanently removed from nuclear waste, all long-term repository concerns would completely disappear.

I indicated earlier that P-T is not a new idea. It has been studied and then restudied every ten years or so. It is usually discarded because it is more expensive than the direct repository disposal process. Its most recent resurrection is due to three principal driving forces: 1) there is growing concern that the first repository may not be built, 2) new chemical technology has been developed to allow partitioning to be done much more precisely than in the past, and 3) the ability to transmute (literally destroy) all the long half-life radionuclides in the two groups discussed above has been shown to be feasible.

The principal attraction of the P-T concept is that it could substantially improve the efficiency of the first repository. In fact, it could possibly even remove the need for a repository entirely. I should hasten to say that the latter statement is highly controversial within the technical community. Hence, I emphasize the word "possibly." Nevertheless, substantial interest has recently developed in Japan, Western Europe, and in parts of the United States. The biggest liability to this proposal is the need to conduct the research and development necessary to streamline the system. Also, it would almost certainly cost more than the current plan.

In the end, the public must decide how much it is willing to pay to solve the nuclear waste dilemma. As we have tried to outline in this chapter, the problem is actually fairly minor from a technical point of view. The amount of waste is small, and the waste eventually decays away to a totally benign residue. The health risks to the public are incredibly small. Hopefully, a wider understanding of these facts will place us in a better position to choose a nuclear waste disposal approach with which we can all be comfortable.

7

NO MORE BOMBS, PLEASE

Proliferation and Terrorism

A Skeptic's View

Well, I'll have to admit that a nuclear energy plant may not be as bad as I once thought, as long as it is used to make electricity, but what about the nuclear bomb? I know for a fact that reactors are an integral part of the bomb making process. We've used reactors for such a purpose in our own country.

The stronger our commitment to nuclear energy, the higher the likelihood that other nations will get the bomb. What would have happened in the Gulf War if Saddam Hussein had actually had one?

Then there is the terrorist threat. Can you even begin to imagine the consequences if a terrorist group stole enough plutonium to make their own bomb?

> *No, even if I could really believe everything in the pre-*
> *ceding chapters, the close association of nuclear energy*
> *and nuclear bombs rules out the nuclear option in my*
> *mind. We simply can't risk nuclear holocaust!*

Nuclear weapons *are* devastating. Other than natural phenomena such as earthquakes, hurricanes, tornados, or volcanic eruptions, they are unquestionably the most destruc-
tive force known to humankind. Consequently, if a new
country wants to develop the bomb or a terrorist group
attempts to build or steal one, we must take all prudent
steps necessary to prevent such an occurrence.

We must also recognize, however, that the link between
nuclear energy and nuclear bombs is nowhere near as
tight as some would like us to believe. Certainly there is a
connection, just as there is between the fire in the backyard barbecue and the fires that have raged through San
Francisco, Chicago, and Yellowstone National Park. The
basic laws of physics and chemistry don't respect the purposes for which they are used. However, we can't put the
genie back in the bottle, so the principal question can be
stated simply. Would nuclear proliferation or nuclear terrorism on the global scale be reduced if the United States
were to terminate its commercial nuclear energy program?

NUCLEAR PROLIFERATION

Who wants the bomb? This question may strike some people as rather strange, because it may seem obvious that
any nation would be better off as a "have" than a "have
not." Actually, it's not that simple, and history has borne
this out.

There is clear evidence that a civilian (non-military)
nuclear energy program does not necessarily lead to a
nuclear weapons program. Currently 36 nations have

nuclear energy, yet only six nations have publicly admitted to having the bomb.[1] In fact, despite much speculation to the contrary, there has not been a new member of the acknowledged "nuclear weapons club" in nearly two decades.

It is important to understand the incentives a nation may have for acquiring nuclear bombs. The two principal incentives are either military or political gain, or some combination of the two.

In the case of the United States, the incentive was clearly military. We were caught by surprise at Pearl Harbor in the Pacific, and the war in Europe was exacerbated by rumors that German scientists were well on their way to building the atomic bomb. The thought of nuclear weaponry in the hands of Adolph Hitler was enough to give the wartime "Manhattan Project" (the code name for the project to build the atomic bomb) the highest priority on the president's agenda. Military stature was also the principal driving force for the Soviet program, which resulted in the detonation of their first nuclear device in 1949. They undoubtedly felt the need for a nuclear counterforce.

It is likely that political gain was the principal driving force for the next four countries to acquire nuclear weapons: Britain (1952), France (1960), China (1964), and India (1974). Britain and France were surely aware that they didn't need the bomb for military protection; the umbrella protection provided by the United States was sufficient to deter aggression. However, both nations had enjoyed a long history with a self image as world leaders, so it is not hard to understand their feeling that it was necessary to be a member of "the club" in order to retain that image.

China may have had mixed motives. Certainly acquiring the bomb gave them new political clout, but a deepening resentment toward the Soviet Union provided appreciable military incentive. India presents the greatest mystery. They have always maintained that their nuclear test explosion in 1974 was for "peaceful purposes." However, there are many who believe this was a deliberate attempt on the part of a Hindu nation to "flex muscle" to their surrounding Muslim neighbors.

Perhaps the most important question is why isn't the list increasing? There must be disincentives to acquiring the bomb. We need to understand these disincentives in order to get beyond the perceived connection between peaceful nuclear energy and nuclear proliferation.

The two principle disincentives are politics and economics. It is interesting that "politics" can be both an incentive and a disincentive. Germany and Japan are the prime examples. There is no question that they have the technology and the infrastructure to develop nuclear weapons if it were in their interest to do so. However, the political pressure in the post-World War II era from the victorious Allies provided enormous disincentives for Germany and Japan to take this step. This, combined with strong internal political restraints, continues to deter both countries.

The economic disincentives are also very real. Building a successful nuclear warhead requires enormous effort. Acquiring the necessary fuel requires either enriching uranium (to concentrate the fissile U-235 species) or using specially designed production reactors (to generate relatively pure Pu-239 from uranium) plus associated chemical plants to extract the useable fuel. In addition, a very sophisticated capability must exist to assemble the bomb. This requires specialties in chemistry, nuclear physics, hydrodynamics, explosives, electronics, materials technology, metallurgy, and elaborate equipment to machine complex shapes. Finally, to have an effective nuclear weapon, it is necessary to develop an elaborate delivery system. The entire enterprise is so expensive that it can only be accomplished by dedicating a significant portion of the total economic resources of a nation.

In fact, there are very few countries in the world where the net political gain, combined with the staggering expense of developing such a capability, would justify entry into the "nuclear weapons club." Yet there are some, and that is enough to cause a great deal of worry. There are still the Saddam Husseins of this world who would likely be willing to pay any price, despite the hardships inflicted on their country's citizenry, to be the bully of the neighborhood. It is precisely because of this that an enormous amount of attention has been quite properly directed at nonproliferation efforts.

Nonproliferation has been one of the principal objectives of U.S. foreign policy since the end of the Second World War. The United States has participated in four major nonproliferation efforts since its inaugural detonation of the world's first nuclear device.

The first era lasted from 1946-1953 and was a period of secrecy and denial. The second era, from 1953-1968, was launched by President Eisenhower's Atoms for Peace initiative. During this period, the International Atomic Energy Agency (IAEA) was created to monitor the flow of sensitive materials in and out of nuclear facilities for those countries that volunteered to participate in international safeguards. The third initiative, started in 1968 and still in effect, is the Non-Proliferation Treaty (NPT), which added further controls. The fourth initiative, enacted in 1978 and also still in effect, is the Nuclear Non-Proliferation Act (NNPA), in which the United States imposed a further set of restrictions.

This latter act, designed and adopted during the Carter administration, was intended to correct some of the loopholes of the NPT. Unfortunately, its initial implementation was conducted in an almost dogmatic manner. Nuclear fuel reprocessing was banned in the United States, and other countries were strongly urged to follow suit. Rather than succumbing to American pressure, several major countries determined that the U.S. was no longer a predictable and reliable nuclear partner and developed their own commercial reprocessing enterprises. France, Britain, and the former Soviet Union now have fully operational reprocessing systems, and a commercial reprocessing plant is under construction in Japan. Several other nations have at least some reprocessing capability.

The technology used by the IAEA to detect any potential diversion of commercial nuclear material into the military sector is, in many cases, used to monitor reprocessing plants. Consequently, the United States, having banned reprocessing development, has lost considerable influence in the very important area of nonproliferation. While the United States' longstanding goal of preventing nuclear proliferation is unquestioned, one could easily argue that the unilateral withdrawal from reprocessing technology development by the United States has been counterproductive to its stated nonproliferation goal. Further, the advantages of reprocessing (discussed in chapter 6) were lost.

NUCLEAR TERRORISM

The other problem deserving attention is the threat of nuclear terrorism. Possession of a bonafide nuclear weapon by a terrorist group would, indeed, be grave cause for concern. Further, the almost paralyzing effect such a group would have because of the great press coverage instantly afforded could convert even the claim of such possession into a powerful force.

How likely is it that a terrorist group could really acquire such a weapon? More to the point, how likely is it that they could build a bomb as a direct result of the U.S. civilian nuclear energy program?

The first major task for a group of terrorists would be to obtain the required fuel. This would have to come from an enrichment plant, a reactor, a reprocessing plant, or from interception during transport. Actually obtaining the fuel from any of these sources would be extremely difficult, since, in most cases, the fuel would be so intensely radioactive that it would be very dangerous to extract.

The security associated with any of these targets is exceptional. Typical barriers include armed guards, high security fences, motion detectors, and remote cameras. Areas containing fissionable materials are further protected by metal detectors so sensitive that in most cases they would detect as little as 0.01% of the quantity needed to make a bomb, even if such quantities were swallowed by a would-be thief. All shipments of such material are made in unmarked armored trucks with armed guards and communication equipment that could readily summon more than adequate reinforcements.

Suppose, however, that somehow terrorists were able to acquire enough weapons fuel to make a bomb. Could they succeed in actually building a bomb? It is difficult to imagine that people with the specialty skills required would normally be an original part of a terrorist group. Consequently, "defectors" would have to be recruited. In addition to the difficulty in attracting the team needed to begin with, one unsuccessful attempt in the recruiting process would destroy the whole endeavor. Furthermore, the machining and assembly of a nuclear weapon is difficult and hazardous, particularly working with the limited facilities that such a group is likely to have.

Tom Clancy portrays such a dedicated effort in his recent, popular novel, *The Sum of all Fears*.[2] However, we must remember that this is a work of fiction. Also, we might note that in his novel, the "defectors" were summarily disposed of (killed) as soon as their usefulness ended!

Given the difficulties facing any terrorist group attempting to build their own weapon, it would seem to be far more likely for a dedicated terrorist group to concentrate their forces on stealing an actual military weapon, already built. This would use the traditional expertise normally associated with such groups (namely, physical force), instead of numerous security penetrations and highly specialized technical capabilities.

Fraught with all these obstacles, it is not hard to see why terrorist groups have used other, much simpler means to accomplish their goals. The 1995 gas poisoning of Japanese subways is a case in point. Other examples include the bombing of the World Trade Center in New York City and the federal building in Oklahoma City, where terrorists used agricultural fertilizer materials to construct their bombs.

IF WE WERE TO GIVE UP COMMERCIAL NUCLEAR ENERGY

We are now in a position to answer the question of whether the commercial nuclear energy industry in the United States contributes to nuclear proliferation or nuclear terrorism. The straight answer is that it contributes very little or not at all. In fact, if we were to eliminate commercial nuclear energy in the U.S., a very strong argument could be made that the chances of nuclear proliferation and nuclear terrorism would appreciably increase.

To address the nuclear proliferation issue, we must return to the reasons a non-weapons nation would or would not wish to acquire nuclear weapons capability. It should be clear by now that a complete shutdown of all U.S. commercial nuclear power facilities would have zero impact on the ability of any other country in the world to obtain the bomb, if they made that goal a national priority. The knowledge is available, and there are plenty of other suppliers. The only leverage the U.S. has is military clout (which would not be affected either way by stopping our commercial nuclear energy program) and political clout. Our political clout would be considerably weakened in the arena of nonproliferation if we are not keeping up with technical advances in nuclear energy. By drifting away from being an international nuclear partner, we would continue to lose influence. This is precisely what is currently happening already as a result of withdrawal policies initiated by President Carter.

By way of analogy, ceasing our commercial nuclear energy program because we fear some tenuous connection to nuclear proliferation is a bit like closing down the American automotive industry because people are being killed in traffic accidents in Europe. Not only would such an action have no influence on reducing traffic hazards along the Rhine, it could well have negative effects, since the American automotive industry is a recognized world leader in developing car safety improvements.

If, on the other hand, the U.S. *accelerates* its commitment to commercial nuclear energy, the only possible negative connection to nuclear proliferation is the exportation of the technology. However, a stronger commercial nuclear energy base in the U.S. would certainly take the strain off coal, oil, and natural gas as means to generate domestic electricity. Consequently, more of the fossil fuels would be available for the developing nations of the world. This could help lower the frustration level of those nations not blessed with natural energy resources, thereby reducing their desire to acquire nuclear warheads as a means to get the natural resources they need.

How about terrorism? By any objective analysis, the chances of nuclear terrorism succeeding must be substantially smaller in a country such as the U.S. than in a developing nation struggling to get into the nuclear business. The security infrastructure in the U.S. has been developed over several decades and, although not invincible, is certainly much more impenetrable than in the developing countries. Consequently, a cessation of commercial nuclear energy in the U.S. would have very little if any effect on the international threat of nuclear terrorism. As in the case of nonproliferation, the loss of U.S. expertise that would result from a voluntary cessation of commercial nuclear energy would reduce the U.S.'s influence to thwart nuclear terrorism on the international scale. To suggest otherwise is to ignore the fact that the U.S. developed the international nonproliferation agreements and treaties from a leadership position.

Dealing with the intricacies of nuclear nonproliferation and terrorism on the international scale certainly provides a major challenge. I recognize that the brief discussion provided in this chapter represents an inadequate basis for understanding all the global implications. However, we must be very cautious of those who deliberately exploit the power of the words "nonproliferation" and "nuclear terrorism" in an attempt to frighten us into abandoning our commercial nuclear energy program. To deny present and future generations the benefits of nuclear technology under such a guise is one of the more tragic forms of manipulation.

8

I DON'T WANT TO TAKE CHANCES

A Responsible Look at Risk

A Skeptic's View

I still have an uneasy feeling that nuclear reactors pose a high risk. Now, I know life can't be completely risk free. After all, I do drive cars, ride on airplanes, and even ski occasionally. But I don't want someone forcing an unnecessary risk down my throat. I don't like the idea of some faceless bureaucrat deciding to build a high-risk nuclear energy plant in my area. I don't want to take the chance.

We seem to live in a very risky world. Hardly a day goes by that we don't read or hear of some new threat to life. We are constantly bombarded with announcements of a few common chemicals that have been found to be carcinogenic or some tragedy across the country that could just as easily have happened to us. It seems as though there is very little that is safe to eat, drink, or breathe—let alone ride on!

But is this really a particularly risky time to be alive? If we look back at the world only a century ago, we find that life expectancy was about 50 years, compared to our current life expectancy of approximately 75 years. Therefore, the sum of all the risks we now encounter must be less than the risks faced by our grandparents. Could it be that many of the large risks we previously encountered have been eliminated, leaving a myriad of smaller but more highly publicized risks?

DAILY RISKS

The assessment of risks needs to be based on measurable scientific data, not on feelings. Though we don't always like to talk about it, the fact remains that all of us will eventually die. Benjamin Franklin was right when he said, "In this world nothing is certain but death and taxes." The question, then, is how and when do we die?

Numerous books, articles, and reports have been written in an attempt to bring risk into perspective. For me, the most understandable approach is that developed by Professor Bernard Cohen.[1,2] He has defined a term called the "Loss-of-Life-Expectancy" (LLE). The LLE is the average amount by which one's life is shortened by encountering a particular risk.[3] I've chosen to present much of his work in this chapter, reformatted in graphical form.

Every time I give a public presentation dealing with the pros and cons of nuclear energy, I ask the audience to tell me what they believe to be the highest risks in life. The answers are completely predictable: driving cars, smoking, nuclear energy, and so on. I then show them information of the type contained in figure 23.

As startling as it may seem, the highest risk to which men voluntarily submit themselves is to remain unmarried. Needless to say, this always draws chuckles and a few clever jokes, at least from the male contingency, and the females invariably sit a bit higher in their chairs. Yet the facts unequivocally indicate that the life spans of both men and women are considerably shortened by remaining single—by about 3,000 days for men (over 8 years), and 1,600 days (about 4.5 years) for women.

Next on the list is smoking. Though the risks of this activity have been highly publicized in the United States during the past decade, smoking is still causing the premature death of millions of people. Male smokers lose about 2,590 days of life, and female smokers lose about 1,530.

Alcohol consumption causes cirrhosis of the liver and contributes to motor vehicle accidents. Note that we lose about a half year from our normal life expectancy just by driving or riding in automobiles, although we can reduce this significantly by riding only in midsize or larger cars or by driving more slowly.

Similar data are available for occupations (figure 23-B). We note that the highest LLE is for those living in poverty. Data provide a clear indication that the populations of wealthy nations have life expectancies decades longer than those of poorer nations. Unskilled laborers live 2.5 years less than both skilled and semiskilled workers. Professional, technical, and administrative workers live 4 years longer, and corporate executives live 7 years longer than unskilled laborers. If categorized by educational levels, the American with a high school education lives about 2 years longer than a grade school dropout, and a college graduate lives over 4 years longer than the person who left school in the early grades.

Coal mining is one of the most hazardous occupations because of accidents and black lung disease acquired from coal dust. On the average, those engaged in coal

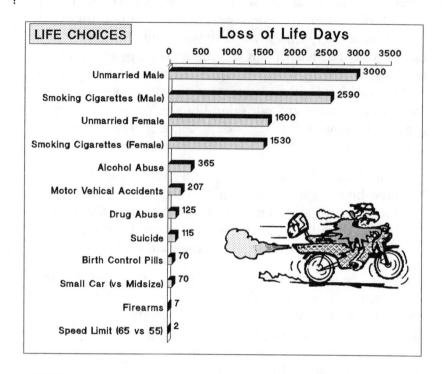

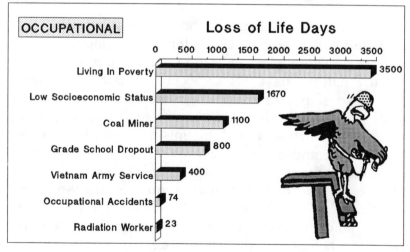

Figure 23-A. Loss of Life Expectancy for Life Choices

Figure 23-B. Loss of Life Expectancy for Various Occupations

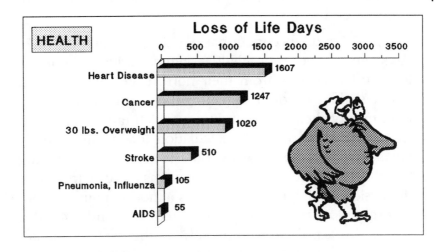

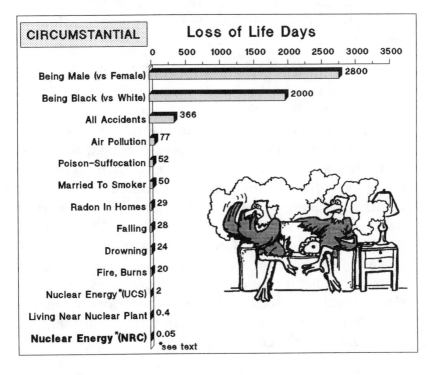

Figure 23-C. Loss of Life Expectancy for Health Ailments

Figure 23-D. Loss of Life Expectancy for Circumstantial Situations

mining relinquish about 3 years of their normal life expectancy.

The LLE for all occupational accidents combined is 74 days (about two and a half months). Radiation workers lose about 23 days from their normal life expectancy, a little less than a month, assuming they are exposed to the full limits of their allowable dose (which rarely occurs). It should be noted that this calculation for radiation workers is based on the linear hypothesis model for estimating health effects from radiation. You may recall from our discussion in chapter 4 that this is almost certainly a considerable overestimate of any actual harmful health effects. In fact, if we were to rely on the data accumulated for the shipyard radiation workers (discussed in chapter 4), we would conclude that the number would be zero, or even negative.

Before we go on, we need to realize that the most dangerous occupation, by far, is no occupation at all, i.e., being unemployed. Statistics indicate that one year of unemployment translates into an LLE of about 500 days. Stated differently, every day of unemployment robs us of more than a day of our life! This is equivalent to smoking ten packs of cigarettes per day. Elevated to a national level, it is estimated that a 1% increase in American unemployment results in 36,000 deaths. This includes 20,000 cardiovascular failures, 500 deaths due to alcohol-related cirrhosis of the liver, 900 suicides, and 650 homicides. In addition to these deaths, there are 3,300 admissions to state prisons and 4,200 admissions to mental institutions. It is clear that increased unemployment has very significant adverse health effects.

Let's move on to consider risks due to health considerations (figure 23-C). As mentioned earlier, heart disease is thought to be the nation's number one killer. Note that the LLE due to heart disease, over 1600 days, is actually less than some of the other factors already considered. As a nation, we have yet to recognize that poverty literally deprives more people of their days on Earth (in addition to the quality of those days) than heart disease.

Cancer probably draws more attention than any other health concern. Indeed, it is one of our leading killers, depriving Americans of over 1200 days of life expectancy. We must be reminded that cancer is a disease principally associated with old age.[4] The older we get, the higher the chances of dying from some type of cancer. Consequently, the primary reason for the higher incidence of cancer in the United States is that, overall, we are dying at an older age.

If our focus is on good health, we may need reminding that most Americans eat substantially more than necessary. In many cases the resulting health effects from such eating patterns exceed those due to cancer. Being overweight by 30 pounds, for example, results in an LLE of 1000 days.

Last, let's look into "circumstantial" risks (figure 23-D). These are the risks over which we have little or no control. For instance, we have no choice in being born male or female, black or white. Yet males and blacks have shorter life spans.

Accidents, too, count as circumstantial risks. Our LLE for all accidents combined is about one year. Accidents such as drowning, falling, poisoning, and fire are all comparable to living with a smoker in the house, in terms of risk.

Here we introduce the risk from nuclear energy. I chose to include it in the "circumstantial" category because many people feel they have no choice in allowing this technology to exist.

As noted in chapter 5, a detailed study on the risk of nuclear energy was conducted by the Nuclear Regulatory Commission (NRC) in the mid-1970s. The NRC is widely viewed within the nuclear industry as very conservative regarding safety matters. They always assume the worst in analyzing any possible accident sequence. For a regulator, such a stance is generally viewed as appropriate. The results of their study, translated into a loss of life expectancy, is that the average American is subjected to an LLE of 0.05 days (about one hour) by the nuclear energy industry. This, by the way, is assuming that *all* of our electricity is generated by nuclear energy, which is about 5 times more than current levels.

Does this strike you as incredible? It's what the data show, even including worst-case scenarios. Let me repeat that **if all our electricity was generated by nuclear energy (rather than the 21% at present), the maximum danger inflicted upon the American public would result in a loss of life expectancy of about one hour!**

Now some may say, "Yes, that's what the nuclear promoters say, but how do I know I can believe them?" Needless to say, there are numerous anti-nuclear groups in our nation that focus on every flaw in the nuclear industry. Of these groups, the Union of Concerned Scientists (UCS) is probably the most critical and vocal. Based in Cambridge, Massachusetts, the UCS boasts of support from a few faculty members at M.I.T. The group has severely questioned the NRC study and has claimed that the real health hazards are higher by a factor of about 40. Consequently, their calculated LLE from nuclear energy is about 2 days (40 x 0.05).[5] Even if their calculations were correct, we see that the risks of nuclear energy are nowhere near as great as the risks inherent in the activities, habits, and problems we take for granted in everyday life.

CATASTROPHIC EVENTS

As we consider the risk comparisons from the previous section, we can see that nuclear energy presents an incredibly low risk. This is true even if one chooses to use the numbers generated by the antinuclear faction.

I suspect one of the things bothering a great many people is the potential for a catastrophe. Dr. Robert duPont, a noted psychologist, has studied the problem in some depth and has concluded that many people become mentally paralyzed with a phobic fear of potential accidents.[6] Such persons become intractably mired with "what if" questions, even if the hypothetical events are totally divorced from reality. Many Americans are much more frightened by one large event that kills a great many people, than other events which kill far greater numbers of people, but do so gradually, over a long period of time.

We see evidence of this all the time. Any airplane crash that kills more than a dozen people is front-page news. Yet well over 1,000 times as many people are killed each year in automobile accidents than die in one of these "headliner" airplane crashes. A classic example is the "disaster" projected by the media of the impending fall of our orbiting Skylab satellite in 1974. It eventually fell harmlessly in small pieces over the upper reaches of Canada. The actual LLE for this event was 0.002 seconds!

Our irrational behavior in this regard should not be lightly discounted. To illustrate this concern, suppose we could invent two technologies, A and B, to produce a certain desired product. Let's assume Technology A is known to cause one large accident per year in which 100 people will die, whereas Technology B will kill 1,000 people per year one at a time (relatively unnoticed). As Dr. Cohen correctly points out, if we choose Technology B to avoid the one catastrophic event per year from

Technology A, we are consciously condemning 900 people to their death. Is it fair that those 900 people must die because we are so emotionally stirred by that one catastrophic event?

Nuclear energy has suffered exactly this mistreatment because of such concerns. If we grant that this phobia of catastrophic events is a reality, it may be constructive to compare the risks of nuclear energy with other catastrophic events.[7] Figure 24 contains such a comparison.

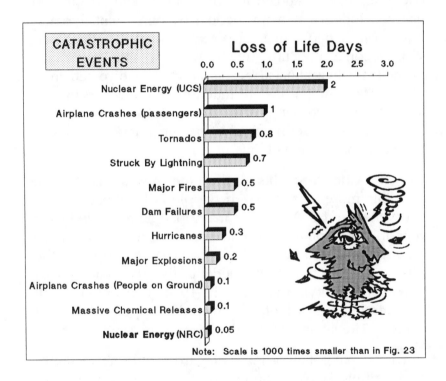

Figure 24. Loss of Life Expectancy for Catastrophic Events

We first need to emphasize to the reader that the scale used in figure 24 is one thousand times smaller than the scales of figure 23, i.e., the LLE for these events is much smaller than the daily risks we discussed earlier. Here we see that airplane crashes reduce our life expectancy by only one day. Yet the media coverage given to airplane crashes has caused many to shake in their seats on every take-off and landing.

The LLE for being killed by a bolt of lightning or dying from other disasters such as hurricanes, tornados, and major fires is about 0.5 days. The 0.05 days for nuclear energy, as calculated by the NRC, is significantly less than these other natural or man-made disasters. We might point out that for all other events listed in the table (i.e., those not associated with nuclear energy), thousands of Americans have actually died. To date, no member of the American public has died as a result of radiation release from the nuclear energy industry. The numbers given are based on "what if" calculations.

NUCLEAR ENERGY IN PERSPECTIVE

Given the data from the above sections, we should now be able to place the risks of nuclear energy in perspective. While every activity of life, including the generation of electricity, entails some risk, the risks arising from not having sufficient energy are considerably greater. Just imagine a hospital without power, and the picture becomes all too clear.

How do other electricity-producing technologies compare from an overall risk perspective? Figure 25 contains such data.

There is an LLE of over 20 days to the American who relies on coal for electricity production, mostly due to air pollution. Such deaths normally come one at a time and are largely unnoticed. Yet every year more than 10,000 Americans die a premature death from this cause. The most notorious public disaster occurred in London in 1952, when air pollution due to coal burning caused

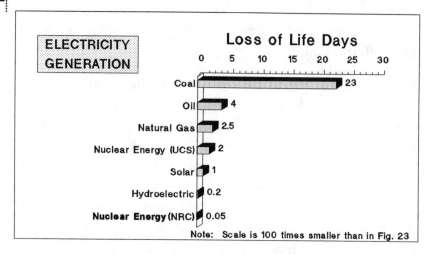

Figure 25. Loss of Life Expectancy for Electrical Generation

3,500 more deaths than normally expected over a span of only a few days.

Oil-based electricity is a little less dangerous, causing an LLE of about 4 days. This number comprises health effects due to both air pollution and fires. Natural gas yields far less air pollution, but numerous deaths have occurred from fires, explosions, and asphyxiation.

Solar energy, should it ever become a major supplier of electricity, is risky from a public health standpoint mainly because of the massive amounts of materials required to build such plants. Energy from some source must be used to refine and manufacture the copper, glass, and other structural materials required. For the calculation shown, this energy was assumed to come from coal. If this energy were to come from oil, natural gas, hydro, or nuclear energy, the LLE for solar power would be accordingly lower. If solar photovoltaics were ever to become a significant contributor to our energy base, there would be additional risks from the hazardous materials used in silicon processing.

We should note that electrocution is a hazard common to all forms of electrical generation and use. The LLE from electrocution to the average American is 5 days. Should solar energy ever become sufficiently economical for it to be generated at individual home sites (as many advocate), the increased potential of electrocution due to homeowner maintenance of such systems would undoubtedly increase the risk of solar energy. Hydroelectric energy, a subset of solar energy, imposes very low risk electricity production, with an LLE of only 0.2 days.

As we said previously, the LLE for nuclear energy (again, assuming all our electricity is generated from this source) is 0.05 days according to NRC calculations. (For completeness, the LLE number of 2 days computed by the UCS is also listed.) We note that coal, the principal competitor to nuclear energy today, has an LLE over 400 times larger than nuclear. Likewise, nuclear energy is 25 times safer than solar power, assuming we choose the NRC calculations for nuclear-generated electricity. No matter how we view these numbers, the actual risk of nuclear energy is incredibly small.

Before leaving this section, it might be instructive to take a broad look at a range of activities that have about the same risk. This list, compiled by Professor Richard Wilson of Harvard University, is tallied in Figure 26.[8] His calculations indicate that each of the activities shown entail an increased chance of death by one part in a million.[9] Although there are some small inconsistencies between these results and those of Professor Cohen (tallied in the previous tables), the results are certainly close enough to be meaningful. All of the activities noted correspond roughly to the lifetime risk that nuclear energy poses to the average American.

The risk of nuclear energy for a lifetime is roughly equivalent to smoking one or two cigarettes, riding in a canoe

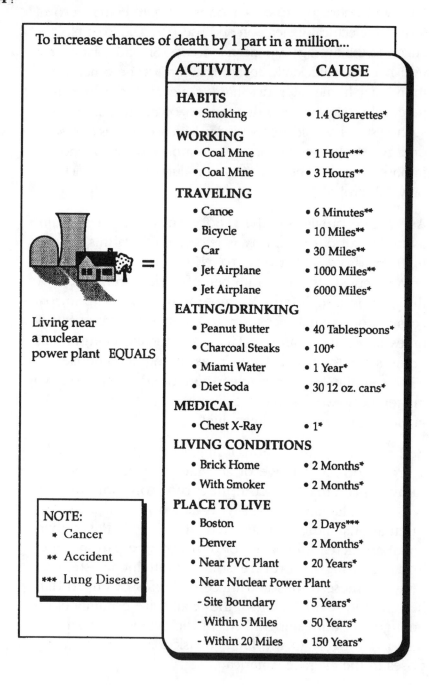

To increase chances of death by 1 part in a million...

ACTIVITY	CAUSE
HABITS	
• Smoking	• 1.4 Cigarettes*
WORKING	
• Coal Mine	• 1 Hour***
• Coal Mine	• 3 Hours**
TRAVELING	
• Canoe	• 6 Minutes**
• Bicycle	• 10 Miles**
• Car	• 30 Miles**
• Jet Airplane	• 1000 Miles**
• Jet Airplane	• 6000 Miles*
EATING/DRINKING	
• Peanut Butter	• 40 Tablespoons*
• Charcoal Steaks	• 100*
• Miami Water	• 1 Year*
• Diet Soda	• 30 12 oz. cans*
MEDICAL	
• Chest X-Ray	• 1*
LIVING CONDITIONS	
• Brick Home	• 2 Months*
• With Smoker	• 2 Months*
PLACE TO LIVE	
• Boston	• 2 Days***
• Denver	• 2 Months*
• Near PVC Plant	• 20 Years*
• Near Nuclear Power Plant	
- Site Boundary	• 5 Years*
- Within 5 Miles	• 50 Years*
- Within 20 Miles	• 150 Years*

Living near a nuclear power plant EQUALS

NOTE:
* Cancer
** Accident
*** Lung Disease

Figure 26. Comparison of Activities with the Same Risk

for six minutes, eating a 24 oz. jar of peanut butter, drinking Miami water for a year, having a chest X-ray, living in a brick home for two months, or living in Boston for two days. The principal cause of death for the examples noted are lung cancer (cigarettes), accidents (canoe), liver cancer caused by aflatoxin B (peanut butter), cancer caused by chloroform (Miami water), radiation (chest X-ray), radiation (brick home), and air pollution (Boston).

A study was conducted in 1982 of the League of Women Voters and a group of college students in which each respondent was asked to rank several activities in the order of diminishing risks.[10] For both groups, nuclear energy ranked highest, well above smoking and motor vehicle accidents. Hopefully, the data presented in this chapter will help all of us make more informed assessments.

CAN WE REDUCE RISKS EVEN FURTHER?

After pondering the data presented above, some readers may be asking whether we can reduce the risks of life even further.

The answer is yes, we certainly can—with enough money. This answer may sound a bit crass, even callous. I can hear the objections now. "There is no way to put a price tag on a human life!" Yet whether we are conscious of it or not, we do put on such price tags, both individually and as a society, and we do it almost daily.

Figure 27-A illustrates the amount of money required to save a human life in the medical field. The numbers are given in thousands of dollars per life saved.

Figure 27-B addresses traffic safety, something we all can relate to. By looking at this list, it is evident that we can save more lives on our nation's highways if we choose to spend the money to do so.

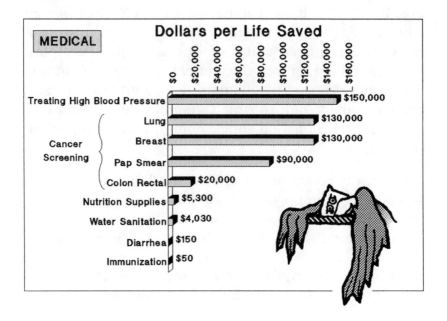

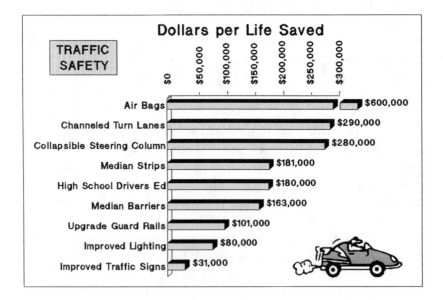

Figure 27-A. Dollars Per Life Saved in the Medical Field
Figure 27-B. Dollars Per Life Saved in Traffic Safety

Finally, figure 27-C shows the costs per life saved in activities involving radiation. Notice that the word "radiation" causes a sizeable step increase in the cost. The numbers reported in figure 27-C are millions of dollars per life saved, 1,000 times larger than for the previous graphs.

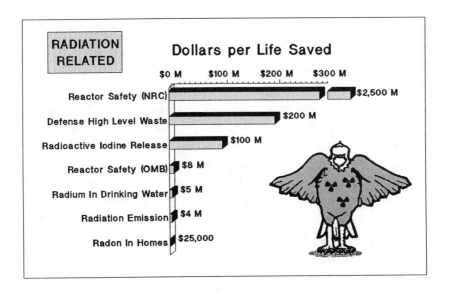

Figure 27-C. Dollars Per Life Saved in Radiation Fields

Reducing radon in our homes can be accomplished fairly inexpensively. The cost per life saved is about $25,000 (comparable to colon screening procedures or improved traffic safety signs). The first step in this process is to implement a program to measure the radon level in homes, which can be done for only about $15 per home.

Standards issued by the Nuclear Regulatory Commission (NRC) result in a price of $4 million per life saved for nuclear plants to meet radiation emission regulations. A similar price tag exists to meet drinking water standards of radium levels. The release of radioactive iodine into the atmosphere carries a particular penalty, amounting to $100 million per life saved.

Reactor safety can indeed be improved, but the cost is incredibly high. In 1972 the Office of Management and Budget (OMB) recommended that nuclear reactor safety systems be installed when they could save a human life for every $8 million spent. Following the release of the NRC safety study mentioned several times earlier, a program was undertaken which resulted in $2 billion to avert 0.8 calculated deaths. Hence, the cost per life saved is a whopping $2.5 billion.

Early safety calculations to support our nation's high level defense waste repository work indicated increased safety could be attained at a cost of $200 million per life saved. Recent EPA radiation standards for the civilian high level waste repository work imply a cost still 10 times higher. Whether it is a nuclear waste repository or radon in the home, it is the same radiation that poses the potential hazard. Yet we can readily see that our government agencies have imposed a regulatory imbalance by a factor of as much as 100,000 times, depending upon the source of the radiation.

These are incredible statistics. Where is our sense of ethics in all this?

To put these numbers into perspective, let's assume you have just been appointed secretary of the new cabinet post of public safety. Your budget for the year is $10 billion. How are you going to spend the money for good public stewardship?

You could spend all of the money to improve nuclear energy reactor safety. According to the above statistics, that might save four lives if the NRC standards are used (10 billion=4 x 2.5 billion per life saved) or 1,250 lives (10 billion=1,250 x 8 million per life saved) lives by OMB standards. You could insist that the money be used to improve the safety of geologic high level waste repositories. That might save 50 lives over time.

If, on the other hand, the money were used to improve traffic safety, the $10 billion could be put into air bags and the number of lives saved would be 16,700. If it were equally spread between driver safety programs, median strips, collapsible steering columns and channeled turn lanes, it would result in the saving of approximately 43,000 lives (close to our annual traffic death toll). If you preferred to spend the money to reduce health risks, the full investment into cancer screening for either lung or breast cancer could save over 75,000 lives. If even ten percent of it were spent to provide nutritional supplies to the Third World, that small amount alone could save 200,000 lives!

The point should be clear. Whether consciously or not, both private and public funds are now being spent to reduce risk in a most irrational, and in my opinion, unethical fashion. We are doing it because of a misinformed public and/or government, and our ignorance has tragic ramifications.[11] Many public officials, responding to public pressure, have continued their harassment of the nuclear industry to the point that utilities have not placed an order for a new nuclear energy plant for more than 15 years. To keep up with customer demand, utilities have been quietly building fossil-fueled plants. Yet every time a coal-fired plant goes on line, it condemns over 1000 people to a premature death over the life of the facility. These statistics are not calculated on a worst-case basis; they are numbers based on actual deaths.

If our response is that we don't want any new energy plants, we must recognize that our economy will suffer, resulting in many times more deaths, as noted in figure 23-B. Indeed, there are risks associated with nuclear energy, but any reasonable assessment would lead us to conclude that the risks we face without it are many times more severe.

9

BUT I'M AN ENVIRONMENTALIST

Our Stewardship of Planet Earth Reexamined

A Skeptic's View

This is the decade of environmentalism. After a century of insulting the earth with technology and industrialization, we finally have enough attention from national and international leaders to be able to do something about cleaning up the air we breathe and the water we drink. We certainly don't want to give up at this point and place more reliance on the epitome of technological exploits, nuclear energy.

Even though nuclear energy may not pose quite the safety risk I once believed it did, I just can't believe it is compatible with our environmental movement. In light of the major gains being made in cleaning up the earth, how could we even consider initiating a long-term dependence on an energy economy based on plutonium, which we all know to be the most toxic substance known to humankind?

There is no question about the strength of the current environmental movement. The U.S. presidents of the 1990s, representing both political parties, have declared this to be the decade of the environment. It's a popular concept that few would dare to challenge.

The reasons for such enthusiasm and commitment appear obvious. During recent decades, many of us have come to realize that there is more to life than material goods and services. The joy of breathing fresh air, drinking pure water, and experiencing the beauty found in undisturbed nature is precious and must not be sacrificed. I am an avid camper and backpacker for precisely these reasons.

Regretfully, all of us have observed the insults that modern life has inflicted on us—smog, belching pulp mills, "aromatic" cattle feed lots, and polluted streams and lakes.

But is technology the real culprit? Has the word "contamination" taken on a new meaning? What is the basis for environmental questions of global significance, such as the greenhouse effect and acid rain? Has the "environmental movement" gone too far?[1]

As we ponder these questions, we need to revisit our options for generating electricity, and determine which approach causes the least environmental pollution at a reasonable cost. Given this perspective, it is then important to allay the near-phobic fears associated with plutonium.

TECHNOLOGY AND THE ENVIRONMENT

As we look at our present surroundings, are we correct in denouncing technology as the enemy of the environment?

Few people alive today remember the odor of horse manure in the city streets, or the smell of human sweat from week-long unwashed bodies. How many can still

recall the musty kerosene lamps and coal fires, the backyard lye pots and the boiling laundry tubs? Without electricity, food was kept in an icebox (for those with access
to ice). Food poisoning was frequent because of the lack
of preservatives and poorly understood methods of safe
home canning techniques. Infectious diseases were a fact
of life. It was a rare family that escaped the ravages of
measles, mumps, pneumonia, diphtheria, whooping
cough, chicken pox, and the like.

Somehow we have forgotten what life was like before
the conveniences afforded us by modern technology. The
quality of life that most Americans now enjoy—resulting
from such technological developments as the widespread
availability of electricity, the extensive chemical and medical advances, rapid modes of transportation, etc.—is
truly the envy of the world.

Yes, you may say, we now enjoy better lives than Gram
and Gramps had, but what are the long-term costs of the
pollution from the pesticides and herbicides that have
made our land more productive and our lives easier?

What about the waste from the electricity producing industries?

Is pollution really worse than a few decades ago? Let's take a closer look. We now have electronic equipment that can measure contamination in parts per million, parts per billion, and even parts per trillion. Could it be that we now have the luxury of worrying about rather insignificant hazards that at one time, even if they could have been detected, would have been given low priority in relation to the much greater hazards of the day?

A classic manifestation of our hypersensitivity to contamination was the alarm registered in the Western world following the Chernobyl accident. I vividly remember reports from Portland, Oregon, claiming that it was among the first sites in the United States to identify drinking water that had been contaminated by fallout from Chernobyl. The original *Oregonian* article, "Portland Radiation Toll Seen," indicated that 654 picocuries of radiation per liter had been found during the peak day, with rapidly falling levels thereafter. Some people undoubtedly thought the sky was falling. However, a picocurie is only one trillionth of a curie (i.e., one millionth of a millionth). Consequently, each Portland resident would have had to drink several hundred gallons of that water *daily* before any perceptible health effects due to radiation could have been detected. As an additional note of perspective, the typical human body contains about 4,000 picocuries per liter at all times (due principally to Potassium-40).

Just how much is one trillionth of anything? To begin to get a grasp on such a tiny number, let's first try to visualize a much easier number, say one millionth. Suppose we were to place basketballs end to end across the U.S., a distance of approximately 3,000 miles. Now assume every millionth basketball is painted red. How many such basketballs would we find along this "basketball highway" across our nation? The answer is about 20.

What if we now sequentially replaced those basketballs with end-to-end tennis balls, ping pong balls, and finally #4 shot BBs? As noted in figure 28, if we painted every millionth #4 shot BB red, we would find about 2,000 red BBs in such a 3,000 mile string.

But now let's paint every billionth #4 shot BB red. How many painted BBs would we find? Only two. Finally, to get back to our original quest, let's paint every *trillionth* #4 shot BB. Now how many red BBs would we encounter? Very likely *none*. In fact, from a statistical standpoint, we would have to line up a string of #4 shot BBs that would circle the Earth some 60 times before we could expect to find a single red "contaminate."

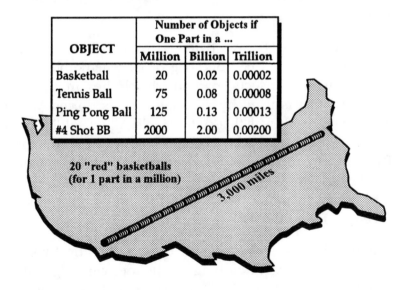

OBJECT	Number of Objects if One Part in a ...		
	Million	Billion	Trillion
Basketball	20	0.02	0.00002
Tennis Ball	75	0.08	0.00008
Ping Pong Ball	125	0.13	0.00013
#4 Shot BB	2000	2.00	0.00200

20 "red" basketballs
(for 1 part in a million)

3,000 miles

Figure 28. Number of "Red" Objects in an End-to-End String from Los Angeles to New York City

It is amazing that modern science has the ability to measure "contamination" in such small amounts. Although technology allows us to detect such infinitesimal contaminants, we should not conclude that there are any health risks involved. The next section provides a brief sample of the unfortunate consequences possible when we make this error.

THE ENVIRONMENTAL MOVEMENT

For most Americans, the mere mention of the word "environment" in a title, campaign, or cause will almost certainly muster support. We all want to improve our environment, but we need to be aware that any word that contains so much political power is subject to abuse. We must be aware of this abuse if we really want to improve our environment, rather than blindly following the agenda of special interest groups. The DDT and Alar stories, two among many, make that abundantly clear.[2]

The insecticide DDT was patented by the Swiss chemist Dr. Paul Muller in 1939. Because of its remarkable effectiveness against insects, particularly clothes moths and body lice, it was used by all Allied troops during World War II. The result was that not a single Allied soldier died of typhus fever (carried by lice), a first in warfare history. (By contrast, more soldiers died from typhus in World War I than from bullets.) Dr. Muller was awarded the Nobel Prize in medicine in 1948 for the medical importance of DDT to the global community.

Perhaps the most impressive measure of the effectiveness of DDT was its use in curbing mosquitoes carrying malaria. Before the use of DDT, approximately 200 million people around the globe were stricken with malaria each year, with about 2 million deaths annually. The numbers dropped dramatically after the initiation of widespread spraying practices in the mid-1940s. Six years after the 1971 banning of DDT in the U.S., which caused other nations to follow suit, the numbers skyrocketed to 800 million cases of malaria and 8.2 million deaths from this disease per year.

More specific numbers for Sri Lanka (formerly Ceylon) are particularly striking. In 1948, prior to the use of DDT, there were 2.8 million recorded cases of malaria. A continual spraying program reduced this to only 17 cases in 1963. Low levels were maintained through the mid 1960s, but then officials became worried about the "environmental effects" because of the debate on this issue in the United States. Sri Lankan officials subsequently discontinued their spraying programs, and immediately thereafter, in 1968, the number of malaria cases soared to 1 million. By the next year, 1969, the number had climbed to 2.5 million, about the same as the pre-DDT levels.

How could a total ban on such a remarkably effective insecticide come about?

The activity started in a reasonable manner. During the height of spraying, trace amounts of DDT were discovered in soil, water, and the bodies of numerous living animals such as birds. DDT was so effective against such a variety of insects that many people were using considerably more than was needed. Some controls were certainly in order.

Rather than conducting the studies necessary to provide prudent controls, however, special interest groups went for the jugular. Energized by the lyrical emotions of Rachel Carson's book, *Silent Spring*, zealous environmental groups demanded a total ban.[3] The attack focused primarily on the possible extinction of certain bird species, the inability of DDT to ever leave the environment, and its potential for causing cancer in humans. None of the charges have ever been convincingly substantiated. Bird species in question, fed with DDT to levels 20,000 times higher than allowed in human food, revealed no statistically relevant adverse health effects. DDT usually breaks down in the natural environment and loses its toxicity within two weeks. Nevertheless, human volunteer groups were fed DDT at levels 500 times the maximum residual DDT found in food during the years of peak spraying. No ill effects were found, either then or in the 30-plus years since that time. Finally, scientists determined that no liver deformations in mice could be detected until the animals received DDT at levels 100,000 times higher than any level that could be found in food residue. Consequently, the National Cancer Institute declared that DDT is *not* a carcinogen.

Given the above information, supported by reams of technical documents and testimonials from 150 scientists, the 1971 public hearings board concluded that a ban on DDT was not desirable. Yet only two months later, Environmental Protection Agency chief William Ruckelshaus banned all uses of DDT, unless an essential public purpose could be proved.

Why? Ruckelshaus later admitted that the decision was purely political. It can only be conjectured as to whether he succumbed to the pressure of special interest groups or whether he saw this as a means to establish the power of the EPA. In either case, the decision was devastating, not only because it deprived humankind of an important weapon against disease but also because it gave credibility to "pseudoscience," an atmosphere in which scientific evidence can be pushed aside by emotion and political pressure.[4] We need to take note.

Now what about the Alar scare?

This case is far less important, both in substance and impact than the DDT story. But it shares one vital element—the power of emotion in the name of the environment.

For decades, America has been proud of the apples domestically grown and consumed. But growing and marketing apples is a very competitive and risky business, and growers and distributors are constantly on the lookout for ways to produce larger, crisper, prettier, and better tasting apples.

Consequently, it is not surprising that growers in the 1980s were attracted to Alar, a growth regulator that was demonstrated to keep the apples on the tree longer, thus promoting a deeper red color and a crisper texture. It also made harvesting more uniform and economical.

Soon afterward, the Natural Resources Defense Council (NRDC) began conducting tests on Alar by feeding large quantities to mice. They announced, with considerable fanfare, that Alar causes cancer. The CBS program "60 Minutes" picked up the story, and a national media blitz immediately followed. Grocers across the country pulled apples from their shelves for fear of public reprisal. The apple industry lost over $200 million, and, in many cases, the wholesale price of a box of apples with Alar dropped below production costs. Many growers were forced out of business.

Was there any real basis for the NRDC's accusation? Extensive studies have since been carried out to ascertain whether Alar could cause cancer, and these studies have failed to find any credible affirmative evidence. The studies indicate that a person would have to eat 28,000 pounds of apples every day for 70 years to produce tumors similar to those observed in the mice. Furthermore, the studies showed that no tumors at all were produced in mice that were fed half the NRDC dosages. In other words, a person could eat up to 14,000 pounds (seven tons) of apples with Alar per day for 70 years without any risk of cancer at all!

Despite this evidence, the EPA elected to decertify Alar. This action was clearly political, a sheer capitulation to public pressure rather than a reasoned decision based on scientific data.

One of the major tragedies of this episode, beyond the needless economic loss for the apple industry, is that there are new technologies available that could allow the delivery of apples and many other farm products to the consumer in better condition. These are food irradiation techniques. However, even though such procedures have been demonstrated for decades to have no harmful side effects, the mere mention of the word "radiation" is enough to propel groups like the NRDC into spasms. Consequently, our citizens (and millions of people around the globe) are being deprived of well-proven methods of avoiding food spoilage because of fear, which some zealous environmentalists have continually managed to exploit.

GLOBAL ENVIRONMENTAL CONCERNS

Two environmental concerns that have received considerable attention in recent years are the greenhouse effect and acid rain. Both have been linked with the post-industrial age by their relationship with our increased reliance on the burning of fossil fuels. As might be expected, substantial controversy has surrounded these issues, both politically and within the scientific community. Because of their potential impact on life on earth, they deserve special treatment.

The Greenhouse Effect

Perhaps no other environmental cause in history has received the national and international attention that the greenhouse effect has. What began as a scientific curiosity only a couple of decades ago has now become the centerpiece of global environmental concerns. It is the focus of international conferences, legions of scientific articles and books, and even legislative initiatives.

What is the Greenhouse Effect? Simply stated, it is the effect of certain gases (principally carbon dioxide) in our atmosphere that trap energy that would otherwise be radiated away from the earth. This causes the temperature at the surface of the earth to rise.

The same effect is noted from a pane of glass. We build greenhouses that allow direct solar radiation to penetrate inward efficiently, but then trap the lower energy radiation emanating from the plants and soil inside the glass house. Indeed, it is precisely this everyday heating phenomenon that provides the name "greenhouse effect" to the atmospheric heating effect that is becoming such a dilemma.

Except for a short anomalous rise about 300 years ago, the carbon dioxide (CO_2) content in the earth's atmosphere has stayed between 260 and 280 ppm (parts per million) for the last 10,000 years (up to the industrial age). We know this from analyzing air bubbles that have been trapped for hundreds or thousands of years in the Antarctic and Greenland ice caps. As such, global surface mean temperature fluctuations have been maintained at less than 3 degrees throughout recorded history.

So what is happening to change this balance? The concentration of gases known to cause the greenhouse effect is gradually increasing in the atmosphere. By 1958 the CO_2 concentration had risen to 315 ppm, and by 1990 the number was up to about 350 ppm. The rate of rise is currently about 1.5 ppm per year. If present rates continue, the atmospheric CO_2 content could double in about one century.

Where does this increase in carbon dioxide come from? Most of it, about six billion tons per year, comes from the burning of fossil fuels (coal, oil, natural gas, and wood). Another billion tons a year is believed to result from the clearing and burning of forests. The annual sum of all forest areas being denuded globally by humans approximates an area the size of Virginia. Of this seven billion tons per year of carbon dioxide that is spewed into the earth's atmosphere, approximately half is retained. The remainder is absorbed into the ocean or otherwise disposed of in our vast global ecosystem. It is the approximately four billion tons of carbon dioxide that are retained in the atmosphere per year that are believed to be causing the continually rising CO_2 concentration.

Carbon dioxide is not the only gas that causes the greenhouse effect. In fact, there about 20 gases that have been identified as contributors. But with the possible exception of water vapors, carbon dioxide is the largest contributor, currently comprising about 50% of the total greenhouse effect. The next four gases, in descending order of importance, are methane, CFCs (chlorofluorocarbons), nitrous oxide, and ozone.

There is much speculation on what the rise in concentration of the above gases will cause. If the greenhouse gas concentration was the sole factor in determining surface temperature on Earth, a climate change of from 2 to 4 degrees should have occurred. The best scientific data suggest that a temperature rise of only about one degree has actually occurred. This difference has not been explained to the satisfaction of the scientific community.

Is this one degree rise in the global temperature significant? Greenhouse effect expert James Hansen believes it is. Dr. Hansen, of the National Aeronautics and Space Administration (NASA), has noted that three of the hottest years during the last century came in the 1980s, and five of the nine warmest have occurred since 1978. He was so confident that global warming had arrived that he offered to bet all comers that one of the first 3 years of the 1990s would be the warmest ever recorded. He won that bet in early 1991 when data confirmed that 1990 was the hottest year on record.[5]

Skeptics were quick to point out that natural global variations are expected to occur. Data indicate fluctuations of one to two degrees Fahrenheit over the past two centuries. On the other hand, reliable data show a trend of increasing global temperatures since the turn of the century.

I need to point out that there remains considerable controversy among atmospheric scientists regarding causes and effects associated with greenhouse warming. Dr. Dixy Lee Ray points out that natural phenomena need to be considered along with human interactions before deriving conclusions.[6] Such natural phenomena include volcanic eruptions, which spew vast quantities of carbon dioxide and other pollutants into the atmosphere. The atmospheric models presently available are still incapable of properly accounting for all the relevant variables. Dr. Patrick Michaels, a well-recognized atmospheric scientist, has compiled impressive data to support his contention that it is increasing cloud covers, rather than increasing temperatures, that may constitute the basis for any concerns.[7]

So the questions we must ponder are: 1) will global atmospheric temperatures continue to rise, and if so, 2) is this caused principally by human activities? 3) If temperatures do continue to rise, so what?

With respect to the first question, no one can say with assurance that temperatures will continue to rise. Temperature increases over the last century are certainly of concern, but we don't know for sure if we are simply in a normal cycle (prior to a cooling period) or whether we are indeed at the start of an ever increasing warming effect.

In response to the second question, again, we can only guess. As noted above, there are several observations we can point to that suggest increased atmospheric pollution by human activities is unbalancing the natural condition and is directly contributing to the temperature increase. On the other hand, very few scientists would bet their professional reputations on this point, given the weaknesses in our current ability to model and understand our global ecosystem.

Consequently we might be tempted to conclude at this point that we should wait and see. But before discarding this whole topic as nothing more than "scientific curiosity," however, we might profit by contemplating the third question; the potential consequences of global warming.

What might be the global consequences if our CO_2 levels were to double? The general scientific consensus, taking into account uncertainties, is that this would result in a global temperature rise of 3 to 8 degrees Fahrenheit. Three degrees would constitute the largest warming in recorded history. Eight degrees would usher in a temperature regime last experienced on earth during the Mesozoic era, the Age of the Dinosaurs. Hence, it is conceivable that in one lifetime, we could change the earth to a climactic regime not experienced for millions of years. It is a possibility we can't lightly ignore.

Dr. Hansen of NASA believes that a doubling of our CO_2 level would substantially increase the frequency of days per year above 100°F. In Washington, D.C., the number of such hot days would go from 1 day to 12 days per year. In Omaha, it would go from 3 days to 21. In Dallas, the number of days in excess of 100°F would go from 14 to 78 days per year.

Precipitation would generally be higher under such conditions, because more evaporation would take place. However, more of this would be felt near the poles and less near the equator. Some studies have indicated that Midwest soil moisture would be reduced by 40% to 50%. Likewise, the Colorado River and the rivers of northern California could be reduced in flow by as much as 40% to 70%.

Finally, a doubled CO_2 level is expected to result in a sea level rise of a possible one to seven feet. The general effects of a rising sea are inundation of wetlands and lowlands, acceleration of coastal erosion, exacerbation of coastal flooding, an increase in the salinity of estuaries and aquifers, and an increase in flood damage due to the higher water base level for storm surges.

Because of the many uncertainties inherent in such predictions, some may choose to discount much of what has been said above. Yet Dr. Dean Abrahamson, a professor at the University of Minnesota and editor of *The Challenge of Global Warming* (from which I have derived substantial material for this chapter), aptly points out that much smaller changes have already caused substantial social and political outcries.[8] What then are we to expect if even a tiny fraction of the phenomena discussed here should actually occur? During the spring and summer of 1988, North America experienced one of its more severe droughts. Even though it was but a glimpse of what may be in store, the news headlines across America provide a better insight into the impacts than any theoretical study. Sample headlines included the following: "Worst Forest Fires of the Century," "Water Rationing," "Crop Failures," "Insurance Companies Attempting to Default on Drought Insurance Policies," "Reduced Growth in GNP Attributed to Drought," and "Billions Authorized in Drought Aid." The hundreds of deaths attributed to the U.S. summer heat wave of 1995 have not gone unnoticed.

So what should we do about all this? We could take the position that we simply don't know enough about the cause-and-effect relationship to do anything. There is no convincing proof that the greenhouse effect is real.

On the other hand, CO_2 levels are now clearly higher than during the pre-industrial period (at least for recorded history). The dumping of six billion tons of carbon dioxide annually into the atmosphere by the burning of fossil fuel is a relatively new activity. Furthermore, the effects, whatever they may be, are likely to be irreversible. If the greenhouse effect is real, the consequences could be devastating.

Faced with such a possibility, it would seem prudent to make changes where we can. Given the enormous consequences that we may still be in a position to avert, responsible stewardship of our planet requires no less.

The most straightforward step would be to substantially reduce our nation's burning of fossil fuels. This is uniformly accepted as the principal way to reduce human contributions to global warming. We certainly can do this. Neither solar nor nuclear energy emit "greenhouse gases." As we saw earlier, solar energy is not a meaningful reality at present. Even if it ever becomes so, it will need a major backup supply to provide steady base power. Consequently, nuclear energy looks particularly attractive.

As a case in point, France responded to the 1973 oil embargo by making a national commitment to nuclear energy. The French increased the nuclear share of their national electrical output from 20% in 1980 to 70% in 1987. By displacing fossil fuel plants over that same time period, they were able to reduce carbon dioxide emissions from 82 million tons to 13 million tons per year.[9] We recognize that this reduction of 70 million tons of CO_2 into the atmosphere represents only about 1% of the annual global CO_2 discharge. However, France is only one country, and the example it has set could have major global effects if followed by other countries.

The next step would be to encourage the move to an all-electric economy. The transportation sector must be weaned from petroleum at some point. It may be that the visible damage to the environment caused by the use of fossil fuels may have more influence in shaping policy than the scarcity of supplies. The use of hydrogen in the transportation sector is increasingly well researched. It is particularly attractive from an environmental standpoint because the combustion of hydrogen simply produces water. Energy will, of course, be needed to produce the required hydrogen, but if we consider the French experience, we again see the valuable contributions that nuclear energy could make.

A third action would be to discourage the denuding of forests. In the tropics alone, deforestation is proceeding at an alarming rate, stripping an area the size of the state of Maryland each year. In our haste to discourage such deforestation, however, we might be reminded that our goal should not be simply to preserve old forests. Indeed, there are good reasons for such preservation, but reversing the greenhouse effect is not one of them. If young trees are planted, they can remove five to seven tons more CO_2 per acre than aging trees. As a result of an aggressive reforestation program, the annual wood growth in the United States is now three times larger than in 1920. The lesson is to plant new trees.

Acid Rain

The other major energy-related environmental concern to come to the forefront in recent years is acid rain. This concern has been particularly acute in the U.S. and Canada, where many believe that a growing number of lakes will no longer support fish life. They attribute this directly to atmospheric pollution from sulfur and nitrogen oxides discharged by industrial plants. One Canadian report claims that nearly 14,000 Canadian lakes are now "dead," i.e., too acidic to support fish life.[10] The Science Council of Canada reportedly fears that another 10,000 to 40,000 lakes will soon die unless acidic depositions are reduced. Another Canadian report claims that over 400,000 square miles in North America now receive rain 10 times more acidic than "clean rain."[11] This represents an area equivalent to 10% of the land mass of the United States.

Similar cries are coming from environmentalists concerned about the reduced fish life in the Adirondack lakes of New York and streams in the Appalachians. Beyond wildlife, people are also concerned about the trees and vegetation that make up pristine areas. Europeans have long been concerned about the health of some of their forests, such as the Black Forest of Western Germany. They have conjectured that acid rain may be a major culprit of the observed blight.

The term "acid rain" is commonly applied to all forms of atmospheric deposition of acidic substances. This includes rain, snow, and dry particles. Acids form when certain atmospheric gases, principally carbon dioxide, sulfur dioxide, and nitrogen oxides, come into contact with water in the air or on the ground and are chemically converted to acidic substances.

We know from earlier discussions that there is a substantial amount of carbon dioxide in the atmosphere. When dissolved in rainwater, this carbon dioxide is converted into a weak acid called carbonic acid (like carbonated water or soda water). As such, all rainwater is slightly acidic.

Where does acid rain come from? Although substantial scientific controversy still surrounds this issue, there is no question that the sources are both natural and human. The question is, which is most important?

In reviewing the scientific literature to date, Dr. Ray notes the general consensus that acid rain is derived roughly 50-50 from natural and human origins, although natural sources vary widely from year to year.[6] Sulfur can enter the atmosphere from the decay of organic matter in swamps and wetlands, or from fumaroles, hot springs, or even ocean spray. A very large amount is believed to be added by volcanic eruptions. On a worldwide basis, volcanoes may spew about 100 million tons of sulfur compounds into the global atmosphere annually. The principal natural source of nitrogen compounds in the atmosphere is nitrogen fixing (resulting in the creation of nitric acid) during lightning storms. Some scientists claim that lightning alone creates enough nitric acid to keep rainfall substantially acidic.

Contamination by humans is also very much in evidence. Despite some conflicting reports, there is a consensus that rainwater is becoming more acidic, at least in some areas. The principal culprit in the United States is believed to be the burning of fossil fuels, especially coal. The amount of sulfur in coal beds varies considerably from region to region but unless expensive atmospheric scrubbers are used continually, prodigious amounts of sulfur dioxide leave the stacks of coal burning plants. According to the United Kingdom's Watt Committee on Energy, a large coal-fired electrical generating plant (sized to provide electricity to a city of about a million people) discharges about 120,000 tons of sulfur dioxide into the atmosphere annually.[12] This is about 250 pounds of sulfur dioxide per person served by that plant per year. Although efficient scrubbers could reduce this somewhat, the coal-fired plants in the U.S. could add about 20% as much sulfur dioxide to the atmosphere as the combined yearly volcanic eruptions world-wide.

Coal burning also produces large amounts of nitrogen oxides, because of the high temperatures of combustion in air that naturally contains 79% nitrogen. The same British study reports that a large coal electricity generating plant produces and discharges about 20,000 tons of nitrous oxide annually.

What is the environmental damage being done by acid rain? Again, the answer to this question is riddled with conflicting reports. In chapter 5 of *Trashing the Planet*, Dr. Ray challenges many of the reports claiming that the loss of fish life is a direct result of increased acidity in rainfall.[6] The large number of variables, such as soil type in the lake beds, measurement techniques, naturally occurring atmospheric changes, etc., render conclusive statements difficult. Any serious student or policy maker would do well to ponder her arguments before leaping to expensive solutions.

Despite the uncertainties, it should be clear from the numbers that considerably more contaminants leading to the formation of acid rain are being dumped into our atmosphere by human activity now than before the industrial age began. Consequently, it should not be surprising that many groups, such as the National Academy of Sciences, have sounded their warnings.

Concerns over the health of plant life, particularly forests, continue to grow amidst a controversy of whether acid rain is the principal cause. West Germany attracted international attention during the 1980s when it encountered a serious blight on trees. The blight spread from 8% of all West German trees in 1982 to 52% in 1987. This caused considerable concern.

So what should we do? Some would say there is no compelling reason to do anything about the acid rain situation. After all, much of it is due to natural causes that have been here since well before recorded history. Furthermore, the whole situation is so complex that it is not clear that acid rain is responsible for as much environmental damage as many would claim. The logical conclusion would be to continue to study the problem and legislate action (such as mandatory reductions in sulfur oxide emissions) if and when the data are irrefutable. There is certainly some logic to this point of view.

On the other hand, it is clear that fossil fuel burning (coal, in particular) is adding appreciably in acidifying global rainfall. While the biological consequences of such increased acidity are admittedly not conclusive, I've never heard anyone claim that this trend is beneficial. Consequently, it would seem prudent to do whatever we reasonably can to wean ourselves from those activities that are known to contribute to the problem. In many areas of the country, utilities are being required to install "scrubbers" to reduce the amount of sulfur dioxide entering the atmosphere. This can help, but at best it can only reduce the magnitude of the problem. It cannot eliminate it.

As in the case of the global warming, nuclear energy stands out favorably among the known sources for generating large amounts of electricity. None of the gaseous pollutants associated with the burning of fossil fuel are emitted during operation. It is a completely "smoke-free" energy source. Within this context, the environmental attributes of nuclear energy become even more obvious.

COMPARISONS OF THE ENVIRONMENTAL IMPACT OF ENERGY SOURCES

Our environment is precious indeed, and those who champion causes to keep our air pure and our water pristine are pursuing a noble and worthwhile cause. The challenge is to find optimal ways to achieve such pursuits while simultaneously providing the energy we need to sustain a healthy and productive life.

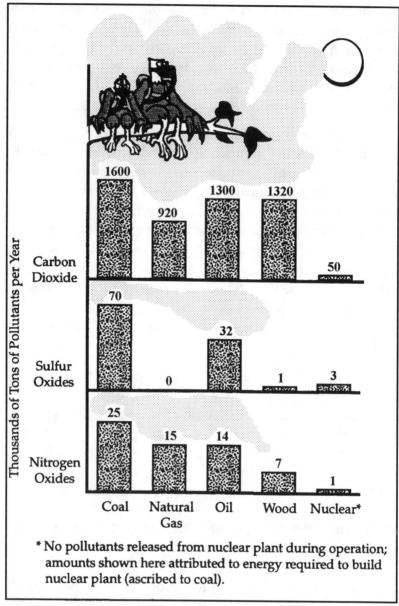

Figure 29. Environmental Pollutants Discharged into the Atmosphere (1000 MW Electrical Plants)

It is instructive to revisit the principal energy production options outlined in Chapter 3 and see how well they stand up from an environmental point of view. Figure 29 provides a comparison of the atmospheric discharges of the three principal gases leading to global warming and acid rain.[13] We see from this figure a rather pronounced difference between the atmospheric pollution generated from fossil fuel or wood burning (biomass) plants and from nuclear plants.

Although not shown in this figure, wood burning is of particular concern with regard to the amounts of carbon monoxide and volatile organic compounds discharged. Nuclear energy releases none of these objectionable gases during operation. The small amounts included on the graph correspond to the pollutants discharged by coal-fired plants, if we assume coal is used to generate the energy required to build the nuclear plant.

The numbers shown for coal burning are lower than those quoted earlier in the U.K. Watts study.[12] That is because the numbers in figure 29 are based on the use of scrubbers, which are effective in reducing the sulfur oxide discharges, but are ineffective for the nitrous oxides or carbon dioxide. It should be noted that coal-fired plants also release several thousand tons of toxic heavy metals into our atmosphere annually. Such elements (specifically listed as objectionable in the Amended Clean Air Act of 1990) include arsenic, chromium, lead, mercury, and nickel compounds. These elements never decay. Because they have infinite half-lives, they stay in the environment forever.

Before moving on, it may come as a surprise to many that sealing our homes and offices as a conservation measure may be more dangerous, from a health perspective, than generating the energy we sought to save. As pointed out in chapter 4, trapped radon gas could lead to numerous premature cancer deaths, if we subscribe to the linear hypothesis model. Trapping atmospheric benzo-a-pyrene likely contributes considerably more than trapped radon to the overall adverse health concerns.

Based on a conservation program initiated by the Bonneville Power Administration in the mid-1980's, calculations indicate an average of nearly 200 potential lung cancers could be developed per year for such conservation measures taken to save the energy from a 1,000 MW electrical generating station.[14] By recalling the vast differences in our sources of radiation as illustrated in figure 17 of chapter 4, we readily see that the comparable number of potential lung cancers from a 1,000 MW nuclear power plant would be well under 1.

A broader environmental assessment would also include direct pollutant discharges to our streams, lakes, and oceans. Oil is particularly notorious in this regard. After the grounding of the tanker *Exxon Valdez* in the Prince William Sound of Alaska in 1989, there are likely few people alive who are not aware of the ecological hazards of transporting oil. More than 10 million gallons of crude oil spilled during that accident, resulting in one of the largest cleanup projects in petroleum history. The January 5, 1993, shipwreck of the tanker *Braer*, pounded by a frothing winter sea in the British Shetland Islands, caused an even larger discharge of oil into the surrounding environment, approximately 25 million gallons. Though normally less dramatic than these two events, spills have become commonplace in the international trade that has become necessary to support our appetite for oil.

In addition to the chemical insults to our environment, thermal pollution is also a factor to be considered. All methods that rely on the production of steam or some other coolant to drive turbines in the generation of electricity must have some mechanism to reject waste heat. By this measure, there is very little difference between the principal fuel systems. All of them operate at about the same thermal efficiencies, which means they all reject approximately the same amount of low quality heat to either the surrounding air or a nearby body of water. It might be noted that while many individuals consider thermal discharge to be an element of pollution, there may be constructive ways to use this waste heat for space heating or other low temperature industrial or agriculture applications.

When placed within the perspective afforded by figure 25 of chapter 8, we note that the overall risk associated with the generation of electricity is quite low, especially as compared to the risks of having insufficient power. Recall the numbers in figure 23-B of chapter 8, dealing with poverty and low socioeconomic status. The point here is that if our goal is to preserve or even improve our environment, nuclear energy looks impressive.

PLUTONIUM TOXICITY

There is one aspect of nuclear energy that has received so much negative press regarding its effect on our environment that it must be addressed. This is the "issue" of plutonium toxicity.

Unfortunately, a few extremists who want to stop the development of nuclear energy have seized on this issue. They have tried to convince the public that plutonium is the most toxic substance on earth. During a speech at Lafayette College in the spring of 1975, Ralph Nader ushered in this myth by stating that one pound of plutonium could kill 8 billion people.[15]

Could it really? Hopefully, by now most astute readers will at least question such sensational statements. Dr. Cohen, the physics professor at the University of Pittsburgh to whom I am indebted for so much of the material found in this book, has studied the toxicity of plutonium for years.[16] In response to Nader's charge, Professor Cohen calculated just how bad plutonium really is.

Plutonium is an alpha particle emitter. As we learned from chapter 4, such particles are stopped very easily by a sheet of paper or a layer of skin. Consequently, ingested plutonium causes very little damage. The problem arises if plutonium is inhaled as a fine dust, because direct damage could be done to the delicate inner surface of the lungs. According to Professor Cohen's calculations, one pound of plutonium completely dispersed as a fine dust in human lungs could theoretically cause 2 million cancers. This is a factor 4000 times less than the Nader conjecture.

This is still a very large number. However, there are several substances that are substantially more toxic. Anthrax spores or botulism toxin, for example, are hundreds or even thousands of times more toxic.

This is not the end of the story. The real question is whether such materials can reasonably be expected to get into human lungs. To answer this question, Professor Cohen hypothesized that a pound of plutonium could somehow be released in the most damaging way in an average big city location under average wind conditions. After taking into account the particle fallout that would occur, the re-suspension of such particles that the wind might cause, and the very long term effects of the plutonium in the soil, the number of potential deaths was determined to be 27. This is nearly a factor of one billion less than Mr. Nader's claim.

We should note that the nuclear industry takes great precautions to prevent the release of plutonium. For starters, plutonium is a solid, not a gas. It is only under extreme accident conditions that it could ever be converted into either a liquid or vapor form. Current EPA guidelines are very strict regarding the amount of plutonium that can be released; namely, no more than one part in a billion of the plutonium handled by a plant can escape as airborne dust. The nuclear industry has always stayed well below these limits.

By far the most important effects of plutonium toxicity are those due to fallout from the atmospheric testing of nuclear bombs. Over 10,000 pounds of plutonium were released to the earth's atmosphere as a fine dust before such testing was terminated in the 1960s. If we were to rely on Mr. Nader's impressions, we would all be dead many times over. Using Professor Cohen's calculations, taking into account the natural geography and population patterns, this amount of plutonium is enough to cause about 4,000 deaths worldwide. This calculation is based on the very conservative linear hypothesis theory outlined in chapter 4. To date, there is no evidence that plutonium has ever caused a human death anywhere in the world—including plutonium workers.

Despite the above evidence, many people continue to argue that one particle can cause cancer. To add some perspective to the highly misleading implications of this statement, Dr. Ray points out that arsenic, cadmium, and chromium are all officially identified as carcinogens.[6] Yet all of these substances are naturally present in every cell of our bodies. How many? Figure 30, which contains the answers, may be surprising.

Such data can hardly be used to support the theory that we need to be concerned about one particle of "highly toxic" plutonium entering our body.

As a good faith measure, Dr. Cohen offered all three national TV networks the opportunity to witness him inhaling 1000 particles of plutonium of any size that could be suspended in air or eating as much plutonium as any prominent nuclear critic will eat or drink of caffeine. The offer was never accepted.

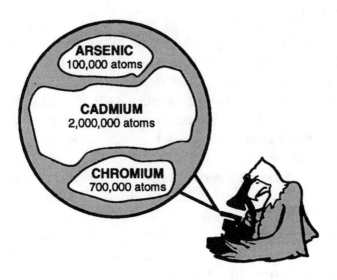

Figure 30. Number of Carcinogenic Atoms in Each Cell of Our Body

So how do we respond to those concerns expressed at the opening to this chapter? Is nuclear energy really the blight that some once thought it to be? Or is it, in fact, an energy source very much in harmony with the goal of achieving energy sufficiency while preserving the health of our planet?

10

I'LL WAIT AND SEE

The Ethics of Choosing Not to Choose

A Skeptic's View

Whew, I need a rest! I didn't realize how many facets there were to consider. I'll have to admit I now have a better appreciation for the issues, but I'm certainly not ready to take up the banner and actually start campaigning for nuclear energy. Well, why should I? Is the earth going to come to a grinding halt if we shut down all nuclear power plants?

Besides, despite what I've just read, I still have a nagging feeling that there is something not right about nuclear energy. I need more time to sort things out. I'll just wait and see what happens.

We can all easily think of several "acceptable" excuses for taking no action. But it is crucial to recognize that choosing not to choose could have enormous consequences. The problem is that examining seemingly controversial sources of energy is more than a mere matter of choice. It is more than simply flipping a coin. There are serious underlying ethical considerations as well. Ethical? Come on, now. Isn't this laying it on a bit thick?

I wish it were. Like many people, there are times when I'd like to close my eyes to vexing problems. The vision of life alone on a remote island in the South Pacific is a fantasy that has admittedly appealed to me more than once! But the harsh reality is that our real island is the planet Earth—and we are not alone on it. In fact, by the time you have finished reading just this one page, another 400 people will have joined us on this "island." The population explosion is nothing less than staggering. We cannot escape the reality of needing to provide an energy support system far more robust than has ever before existed. Without it, there may be no hope for a future worthy of the dignity of human life.

We live in a complex world, and it is growing more complex every day. Isolationist policies of decades past are gone forever. Whether we like it or not, we are becoming a very interdependent people. The recent passage of the North American Free Trade Agreement (NAFTA) and the General Agreement on Tariffs and Trade (GATT) are but two recent examples of the economic ties that interlink the world community.

Given this growing complexity, how do we determine whether any particular technology makes a net positive contribution to society at large? Scientists and engineers can develop a technology, but they cannot provide answers to this question. Such decisions ultimately depend upon the prevailing value system, whether applied at the individual or the collective societal level.

It is within such a context that we are challenged to seriously reflect on how we make decisions on matters of such wide-ranging importance as nuclear technology. Is our ethical framework broad enough to encompass the full range of long-term global implications? Are we willing to honestly probe the consequences of erasing nuclear technology from the face of our planet? Is such a demise realistically possible? Finally, as we ponder the future from this broader perspective, we come to the most basic question any parent can ask; namely, what kind of an inheritance do we wish to leave our children?

THE WORLD WITHOUT NUCLEAR ENERGY

One way to test our ethical framework as it relates to nuclear energy is to speculate for a moment on what life would be like in the United States and the rest of the world if nuclear technology were to vanish. What would our future be like if there were no more nuclear reactors? Figure 31 was prepared to guide our thinking on this important question.

The most obvious initial impact would be *an almost complete dependency of both the United States and the remainder of the developed world on fossil fuel for supplies of electricity.* If we reflect on the U.S. nuclear electricity generation in figure 13 of chapter 3, we soon realize that about 150 to 200 new coal-fired plants would have to be constructed in the United States to generate the electricity now produced by the approximately 110 domestic nuclear reactors. On a world-wide basis, another 300 to 500 coal plants would be required, just to keep even. None of these figures allows for needed growth. If the same nuclear electrical capacity were to be replaced by oil or natural gas-fired plants, the number of plants would be even larger, because these types of plants are generally of smaller size.

Fossil Fuel
Dependence

Less Productive
Agriculture

Environmental
Degradation

Political
Instability

Devastated
Economy

Lack of
Special Medicines

Population Explosion

Figure 31. Implications of a World without Nuclear Energy.

Given this reliance upon fossil fuels by the developed
nations, think of the implications for the third world. If
the United States, or any other wealthy nation, believes it
needs additional quantities of oil to fuel its economy, will
it proceed to secure those supplies? You bet it will!
History reveals no other lesson so clearly. American
forces in the Gulf were not gathered there because of
their intrigue with the Saudi sands.

So what will happen? The U.S., along with other developed nations, will continue to pump the remaining global
fields of oil, while at the same time driving up the prices
of these rapidly depleting supplies.

Where does this leave the third world countries? What energy sources are they expected to use? What are the ethical implications of leaving the developing nations, already mired in poverty, in a position of trying to acquire oil at increasingly high prices, or, worse yet, leave them with no oil (or gas) to obtain at any price? Is our ethical framework large enough to take such factors into account?

The second fundamental implication of a "nuclear free" world is likely to be a stagnant, declining, or possibly even *a devastated economy*. We recall from chapter 1 that there is a clear one-to-one relationship between energy use and our Gross National Product (GNP), particularly between the GNP and the supply of electricity. Insufficient electricity, or electricity at an appreciably higher price, is almost certain to spell economic chaos. Renewable energy, such as solar, is not an economic replacement for nuclear energy. Despite its aesthetic appeal, the basic laws of physics dictate that solar electricity will remain expensive. There will always be a high materials and real estate price tag associated with collecting this inherently dilute energy source. For solar energy to be available in large quantities, we must be prepared to pay the price.

It is also sobering to recall that there are very real health effects associated with insufficient energy supplies. A study conducted at Johns Hopkins University in 1980 indicated that a 1% rise in unemployment in the United States resulted in 36,000 deaths. Every day of unemployment statistically strips the unemployed of more than one day of his or her life expectancy. This is cause for somber ethical reflection.

The third major factor that would eventually be evident in a non-nuclear world is a marked *degradation of our environment.* Increased dependence upon fossil fuels for our energy supplies will result in increased discharges of carbon dioxide and the sulfur and nitrogen oxides that contribute to acid rain. Whether the greenhouse effect is real or not, only time will tell. But do we want to take the chance? Is it ethically correct to subject all citizens on the planet to the possible consequences of global warming if we could instead use a smoke-free energy source such as nuclear energy? Indeed, we must be careful about the way nuclear energy is used to ensure that a recurrence of a Chernobyl-type accident, which did pollute the atmosphere, is diminishingly small. But without nuclear energy, large-scale continued pollution is an absolute certainty. It has been aptly said that pollutants, once in the atmosphere, recognize no political or economic boundaries. Is our value system large enough to factor in these considerations?

A fourth consequence of terminating nuclear technology is its *significant impact on medicine.* Over forty percent of the utensils now routinely used in the medical field are sterilized by radiation techniques. Nuclear medicine is making enormous strides in both diagnosing and treating serious health disorders. Radioactive iodine isotopes are currently used to treat thyroid disease in 30,000 patients a year. A prime example is their use for Graves' disease, which afflicted former first lady Barbara Bush. Over 35,000 procedures are taking place every day in our nation's hospitals that are directly dependent upon the nuclear industry. With the research and development currently underway, these figures should increase dramatically.

Impressive as these figures may be, I must admit that I didn't pay particular attention to this area until I received that fateful telephone call we all hope will never come. My mother called to say she had a lump on her right breastbone, which her doctor believed to be malignant.

For those of you who have ever had to deal with the word "cancer," you know the emotional trauma which immediately sets in. We set up an appointment and took her to a specialist in a larger city. Though hoping to learn of a misdiagnosis by the hometown doctor, we were all braced for the worst.

Shortly after checking into the Seattle clinic, she was injected with a small amount of technetium-99m, a metastable isotope derived from special materials made in nuclear reactors. Upon reading the results, the specialist quickly and positively announced that she had nothing to worry about. She likely had inflamed arthritis. There were no "hot spots" in her bone structure, the telltale indications of bone cancer that would have been clearly revealed by this relatively simple non-surgical medical procedure. Technetium-99m is in such demand that international pharmaceutical companies are actively seeking additional long-term suppliers. There is nothing like personal involvement to drive home the relevancy of such marvelous nuclear medicine advances.

New methods of affixing radionuclides to special molecules that seek out cancer cells, so that radiation treatment can be applied directly to the diseased cells (so-called "silver bullets"), are in advanced stages of development. This would allow highly efficient destruction of cancer cells, without undue damage to nearby normal tissues. The initial therapeutic results on humans are impressive. So prevalent is nuclear technology in medicine that 10 of the last 15 Nobel Prizes in medicine and physiology could not have been attained without radioisotopes, most of which were made in nuclear reactors. What are the ethical implications of denying humankind the medical advances that nuclear technology can offer?

A fifth consequence of eliminating nuclear energy from the globe is continuing the *population explosion*. If present trends continue, the world population will double in 45 years, adding another 5.5 billion people. Reams have been written about the causes of this staggering rise, particularly in the third world. There is no question about the stabilizing effect that ample and affordable energy supplies could render. As Dr. Margaret Maxey has pointed out, the only known approach that will effectively stabilize population is the achievement of moderate levels of prosperity and health for those deprived of it.[1] The poor of the world need food, clothing, and shelter and a kind of security that does not depend upon having more children to provide necessary labor and care of aging parents. As evidenced by the much reduced population growth of the developed countries, there is a minimum level of health and prosperity that only electricity, in reliable and affordable quantities, can provide. There is perhaps no higher ethical goal than seeking ways to reverse the staggering population boom, especially in the poorest of nations where the problem is the most prevalent.

In a closely related area, *agriculture* continues to be a primary beneficiary of nuclear technology.[2] Several improvements in plant breeding have been made possible by using radiation for genetic engineering. For example, a new cotton hybrid, made possible by radiation, has resulted in a medium sized cotton plant that is particularly tolerant to heat and resistant to insects. This has become the most popular cotton variety grown in Pakistan and has allowed production to nearly double in five years, from 4.5 million bales in 1983 to 8.5 million in 1988. Similar results with cotton are being achieved in China and Brazil. Fifty percent of the pasta now consumed in Italy is derived from plants treated by radiation techniques.

As an example of a completely different agricultural application, radioactive isotopes were used in Indonesia to study the digestive systems of buffaloes. This allowed scientists to develop a tailored multinutrient lick block for the buffaloes. Such buffaloes subsequently increased the rate at which they gained weight by over 6 pounds per week. Additionally, the buffaloes produced 1 ton of bodyweight with only 10 tons of grass, whereas it formerly took 50 tons for the same weight gain.

Finally, there is the question of *political stability*. The U.S. flirted with energy rationing in the early 1970s. Even though the energy shortage was minuscule compared to what it would be if nuclear energy were to be eliminated, substantial anxiety was apparent. In fact, isolated shootings were reported in the long lines at the gas pump. Based on this limited experience, there is no question in my mind concerning the political implications of a severe shortage of electricity. Finger pointing and fault finding would be followed by massive political pressures to restore our "rightful" supplies of electricity. Rationing would rapidly evolve from voluntary to mandatory.

Beyond the rapidly crumbling political structure of the U.S. under such conditions, how about the rest of the world? Given our modern communication systems and increasing international travel, there are few inhabitants of our globe who are not at least somewhat aware of the lifestyle that abundant energy supplies have brought to the Western world. It is naive to believe that citizens in any part of our planet will sit still as their access to energy supplies diminishes. The ethical implications of denying our citizens a proven energy source of the magnitude of nuclear energy, and risking the onset of anarchy and even world war, stagger the imagination.

The above observations touch on only some of the obvious implications of a world without nuclear energy. It is possible, of course, that the United States would unilaterally decide to ban nuclear energy, independent of the rest of the world. From a global environmental standpoint, this would not be as disastrous as would be the case if all nations were to abandon nuclear technology, but it should now be clear that at best this would be economic suicide for the U.S.

To get a glimpse of just how large the economic bite in the U.S. would be, two companion studies were conducted for the year 1991 to determine the dollar contributions of the nuclear technology industry to our overall economy. [3,4] Amazingly, the total annual value of this industry was found to be over $330 billion and over 4.1 million jobs. Put in perspective, this is well over twice the size of the largest corporation in America, General Motors. In fact, there are only ten other nations on the face of the earth that have a total GNP exceeding the value of the nuclear technology enterprise to the United States! The current impact on America, which contributes to approximately 300 job classifications, is considerably larger than most people have imagined.

If we are to deny our future generations an energy source based on this technology, I believe we are obligated to think through the implications, based on an ethical framework that takes into account the real needs of our fellow citizens of planet Earth.

COULD NUCLEAR ENERGY BE LOST?

Is it realistic to postulate that nuclear energy might truly die out? Could a multibillion dollar infrastructure that currently produces nearly one-fifth of the world's electrical supply actually cease to exist?

Many people think so. In February 1985, *Forbes* magazine featured an article on the crumbling infrastructure of the nuclear industry. The article provided powerful arguments that this energy technology faces impending death, at least in the U.S.[5] Numerous other writers and politicians have been trying to drive the final nails into the coffin for years.

I would suggest three principal indicators that bear watching as we struggle to see whether nuclear energy can survive: the electric utilities, industry productivity (or lack thereof), and student enrollment in nuclear science at the universities. Even now as it becomes increasingly clear that new power generating capacity is needed, there is probably not a chief executive officer of any utility in the nation who is ready to face the public relation consequences, probable legal suits and costly delays that would likely follow announced intentions to build a new nuclear plant. He or she would literally be "betting the company" on such an order. Why has electric utility confidence in new nuclear energy plants sunk so low?

Are nuclear plants *intrinsically* expensive, relative to alternative means to generate electricity? The answer is certainly no. Essentially all of the early nuclear plants in the United States were cost effective. In fact, most of them produced electricity at consumer rates considerably below the competition. It was only the tax free, federally approved hydroelectric plants, mainly in the Pacific Northwest and Tennessee, that were able to produce electricity at a lower cost. But then the regulatory squeeze came. Largely in response to the accident at Three Mile Island, the Nuclear Regulatory Commission began to mandate tighter criteria, which in turn resulted in numerous redundant safety systems and piles of reporting systems, quality assurance checks, etc. This regulatory squeeze at least doubled the cost of producing power. Protracted public hearings and prolonged litigations all added considerably to construction times. To make matters worse, many state public utility commissions (agencies that control electricity prices) prevented utilities from including construction costs in the electricity rate base before the plant became operational. This forced utilities to borrow all the construction money, often at very high interest rates, or to shift costs to their stockholders.

In order to appreciate what this means financially, let's run through a simple example. Suppose a new power plant were to be built that has a firm cost of 1 billion dollars and a construction schedule of 5 years. If the utility is allowed to raise the price of electricity from their existing plants to provide the funds needed for the new plant, the total cost to the ratepayer is 1 billion dollars. If, on the other hand, the utility is prevented (by the Public Utility Commission) from raising its electrical rates, the money has to be borrowed (normally through issuing bonds). If the interest rate is 15% and the payback period is 10 years, the total cost for the same plant is more than 2 billion dollars. If regulatory requirements or public interventions cause the plant construction time to double (to 10 years), the cost of the plant could again double. Consequently, if this delayed construction plant is financed with borrowed money, the total cost of the new plant is then increased to over 4 billion dollars.

We need to be aware of this elementary arithmetic in order to appreciate the profound differences in the eventual cost to the ratepayer, depending on how the utility is treated in its efforts to build the plant. Those who oppose plant construction can (and have on many occasions) employed these seemingly innocuous devices to drive the price of nuclear plants to many times their intrinsic cost. This is a self-fulfilling prophecy, pure and simple. It is the ratepayer, of course, who ends up footing the bill.

Even given these handicaps, the cost of electricity from American nuclear energy plants today remains about the same as that for coal. It is significantly cheaper than oil-fired plants. Yet during recent years, the only new construction by utilities has been small gas-fired units. These units currently enjoy artificially depressed fuel costs, are relatively cheap to build, and are readily accepted by the public. Given this situation, there is little incentive for a utility executive to fight for the construction of a new nuclear plant.

In marked contrast, nuclear energy in Japan and Europe continues to be very competitive with all alternate sources. The cost of coal plants for new units going into operation in Europe and Japan in the mid 1990s is predicted to range from 24% to 78% higher than new nuclear plants.[6] There is no reason that American utilities can't return to nuclear energy, particularly with the new, more robust designs now being certified. They will only do so, however, if the American public will allow them to proceed.

The most insidious indication of a dying nuclear industry in the United States is the near paralysis that is creeping into essentially all parts of the industry. There is now so much public scrutiny involved in any operation labeled "nuclear" that executives are far more concerned about making a mistake, however small, than about producing anything.

This may seem ludicrous, but this is the reality of the nuclear industry. After reeling from media attacks asserting "production first, safety second," the U.S. Department of Energy has gone out of its way to polish existing procedures, generate new procedures, and dispatch "tiger teams" to ensure a strict adherence to procedures. Media and legal attention to "whistle blowers," i.e., those who claim management has paid insufficient concern to safety concerns, has caused near rigor mortis at many board meetings. This is not to say that the industry should be callous to real safety issues, and it is not to say that mistakes have not been made. It is simply to point out that the pendulum has swung far off center, trapping many company CEOs and bringing productivity to a near halt.

As a case in point, I am acquainted with a fuels specialist employed by a major international nuclear fuel manufacturing firm. One of his responsibilities is to travel throughout the world conducting nuclear fuel examinations as part of an ongoing effort to improve fuel performance. A few years ago he and his colleagues would typically spend one day in training prior to entering a nuclear power plant and then spend another four or five days removing and inspecting the fuel.

Then came the national concern for drug testing for employees in critical jobs (such as airline pilots, nuclear power plant operators, etc.). Now there is nothing wrong with such concerns. Drugs certainly have no business in the workplace, particularly where they could influence the judgment of employees who directly affect public safety. The question is, how far do we take this? Predictably, the Nuclear Regulatory Commission issued a new set of rules stipulating that all employees in critical jobs within the nuclear industry take periodic drug tests. Reasonable enough so far. The problem is that the entrance requirements for each new job now consume four to five days before the job itself can start. The reason? On every new assignment my acquaintance is required to have a urine sample taken for drug testing before any paperwork can start. Further, the sample must be taken on site, at each new site, even if the last one had been conducted as little as a month earlier. Because of the time taken to process the sample, he must either sit and do nothing or take the plane back home and wait for the results before being allowed to start the job. All the check-in procedures are followed to the letter of the law, regardless of cost.

Another example of near paralysis is the difficulty of licensing a repository to accept high-level radioactive waste. Although we certainly don't want to belittle any substantive safety issues that must be resolved as part of the site characterization process, the fact is that the nuclear industry is being held to health and safety standards far more rigid than any industry in history.

This brings us to the final major element of concern, the attraction of the nuclear industry to students. When I decided to go into the nuclear engineering field some thirty years ago, this was the place to be. It had the luster of being a technical challenge and it was known around campus that this was a growing field, which combined good science with the opportunity for enormous public benefit. I felt good about my chosen profession.

Things have changed drastically since then. It is now known around campus that this industry is on the rocks. Students now flock to other disciplines. The number of universities in the United States that still have accredited programs in nuclear engineering are about a third less than only ten years ago, and less than half than when I was in college. Research or training reactors, once considered a vital part of a good nuclear science program, are being rapidly shut down and dismantled. This has happened at two of the three schools from which I obtained degrees.

The full consequences of this reality have yet to be fully felt, but they will be. An industry such as this, which demands technical excellence, cannot succeed without a well-trained and dedicated work force. Unless we decide that nuclear energy is a vital part of our future and provide the proper lure for fertile young minds, our future as a nation is in serious jeopardy.

OUR CHILDREN'S INHERITANCE

So where are our values? Are we now at a point where we are willing to give serious thought to the world we wish to pass on to our children and grandchildren? In so doing, are we willing to take into account the plight of other people's children as well, regardless of their color, culture, or where in the world they happen to reside?

Certainly energy supplies are not the only things in life that count, but it should be clear that the availability of abundant, environmentally acceptable, and relatively inexpensive energy supplies is of monumental importance. So we conclude this chapter and this book by asking whether it makes sense to continue our addiction to fossil fuels, particularly if we have the option not to do so.

I submit, as final graphics, two illustrations I almost always use when making oral presentations on this topic. Both graphics were prepared from the data of Professor M. King Hubbert.[7] Professor Hubbert served many years on the faculty of UCLA, after working for most of his career in the oil industry. Because of his long and painstaking devotion to the field of energy resources and use, he is considered by many to be one of the most credible scientists who ever worked in this field.

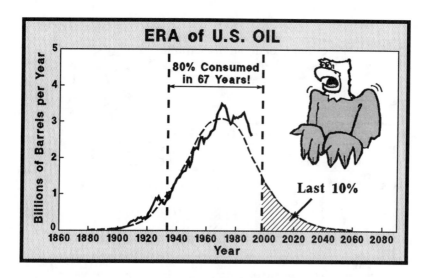

Figure 32. Era of U.S. Oil

Figure 32 illustrates the United States production of oil from the days of the U.S. Civil War up to nearly the present. This is then related to the estimated total producible oil (the area under the "bell" curve). Based on Professor Hubbert's data, and updated to 1991, we see we have now used over 80% of all the oil buried within the 50 states. At our current rates of consumption, Professor Hubbert predicts that we will begin to dip into the last 10%

of this fuel supply within the next ten years. Furthermore, he predicts we will consume 90% of the producible global oil by about the year 2050. This means that most of the world's oil supply will be consumed during the lifetime of our children. Equally unnerving, the longevity of our natural gas supplies is about the same as for oil.

Even more gripping is the information illustrated in Figure 33. This represents the epoch of fossil fuels from the perspective of several millennia, rather than the two to four year election cycles that garner so much attention in American politics. It illustrates the global consumption of fossil fuels from the earliest recorded history up to the present and beyond.

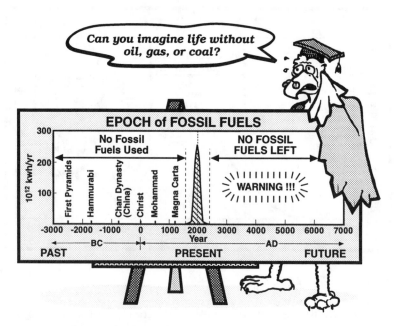

Figure 33. The Epoch of Fossil Fuels

The world supply of oil and natural gas will almost certainly be gone within the first quarter of this narrow epoch. It is only the abundance of coal that could stretch this epoch to three centuries or slightly beyond. If we have any sense of history, or place any value in the welfare of unborn generations, we realize that this time slice in the history of humankind is startlingly stark and narrow!

It is not common to see a scientist overcome with emotion. Yet I'll have to confess that I always have difficulty getting through this final graphic because of the profoundly sobering message it contains. In fact, during one presentation for a Canadian Seminary, the full gravity gripped me to the point of unabashed tears. My eyes are moist as I type these words.

Now Professor Hubbert may be wrong. Even though his predictions for energy consumption have been alarmingly accurate over the past several decades, skeptics point out that the discovery of new deposits of coal, oil, or natural gas have kept extending our fossil fuel reserves. Consequently, it might be a few more generations before our fossil fuels are gone. However, even if this is the case, the point is that it *will* happen. Can you even begin to imagine what life on this planet will be like without oil, gas, or coal? Is this what we want to leave our children?

We still have choices that we can make, but we can't sit back and refuse to become involved. We have the opportunity to employ nuclear energy to a much greater extent than presently allowable, and preserve fossil fuels over a much greater span of time, if we have the vision and will to do so.

Winston Churchill is said to have noted that "Americans always do things right...after they have exhausted all other possibilities!"

Do we have the courage and resolve to allow nuclear energy to develop into an energy and technology worthy of our ethical expectations? Are we willing to do our part to awaken our slumbering eagle before it is too late? Or do we sit idly by and pay the inevitable consequences? The quality of the planet that we pass on to our children and grandchildren awaits our reply.

GLOSSARY

Acid rain—rain (or snow) that contains an abnormally high acidic content resulting from high atmospheric concentrations of carbon, sulfur, or nitrogen oxides; often attributed to the burning of fossil fuels.

Acidity—the range between 0 and 7 on the pH scale (0 is maximum acidity; 7 is neutral).

Actinides—elements with atomic number 90 through 103, which includes uranium and plutonium.

Alar—a growth regulator used by apple farmers to enhance the quality of their produce.

Alpha particle—a positively charged particle made up of two neutrons and two protons (a helium atom nucleus) emitted in certain types of radioactive decay. It can be stopped quite easily; it cannot penetrate a piece of paper or the outer layer of the skin.

Atom—the basic component of all matter. It is the smallest part of an element having all the chemical properties of that element. An atom consists of a nucleus (that contains protons and neutrons) and surrounding electrons.

Atoms for peace—an initiative signed by President Eisenhower in 1954 to allow the peaceful uses of atomic energy to be available to other nations.

Background radiation—radiation arising from natural sources always present in the environment, including solar and cosmic radiation and radioactive elements in the atmosphere, the ground, building materials, and the human body. The background radiation level for an average American is about 370 mrem per year.

Becquerel (Bq)—a measure of the rate of decay of a radioactive substance. One Bq is 1 disintegration per second. (The human body has thousands of becquerels.)

Beta particle—a negatively charged particle (an electron) emitted in certain types of radioactive decay. It can be stopped by a thin piece of aluminum or a short distance in air.

Biomass—trees or other plant material that can be burned to produce energy.

Biosphere—the portion of the earth's surface and atmosphere that supports life.

Breeder reactors—nuclear reactors that generate more useable fuel than they consume.

British thermal unit (BTU)—the quantity of heat required to raise the temperature of one pound of water by one degree Fahrenheit.

Canister—a corrosion-resistant container used to enclose high-level nuclear waste during long-term storage.

Carbon dioxide—a common gas molecule consisting of one carbon atom and two oxygen atoms created by burning carbon substances such as fossil fuels or by metabolism in the body.

Carcinogens—chemical substances or types of energy that can lead to the onset of cancer in the human body.

Chain reaction—a continuing series of nuclear fission events that take place inside a nuclear reactor. Neutrons produced in one fission event cause another fission event.

Chlorofluorocarbons (CFCs)—gases containing carbon and fluorine in various combinations. CFCs are generally quite volatile and are alleged to cause a decrease in the protective ozone layer.

Cladding—metallic tubes that encase nuclear reactor fuel.

Cold fusion—the process of attaining the fusion of two light atoms at essentially room temperature—never yet attained. Many scientists believe it is impossible.

Containment—a heavy structure completely surrounding a nuclear reactor to prevent radioactivity from getting into the atmosphere in the event of a major accident.

Coolant—a gas or liquid used in a nuclear reactor to remove the heat generated by the fission process.

Core—the center fuel region of a nuclear reactor.

Core meltdown—the process of a changing geometry in a nuclear reactor core should melting occur as a result of losing the coolant.

Cosmic rays—naturally occurring radiation that exists in outer space (radiation coming from the cosmos). Many of these are absorbed by the atmosphere.

Criticality—a condition in which a chain reaction involving neutrons and fuel is self-sustaining.

Curie (Ci)—a measure of the rate of decay of a radioactive substance. One Curie is defined to be the radioactivity of one gram of radium—37 billion disintegrations per second.

DDT—a chemical insecticide used to control the population of insects.

Dose—an amount of radiation or energy absorbed (often measured in mrem).

Electron—a basic particle that has a negative electrical charge and has very little mass compared to the nucleus.

Element—an atom with a unique number of protons in its nucleus. Oxygen is an element that has eight protons in its nucleus.

EPA—Environmental Protection Agency, a federal agency chartered to focus on environmental protection activities in the United States.

Fast reactors—nuclear reactors that employ fast (high speed) neutrons to sustain the fission process. Such reactors require the coolant to be a heavy liquid metal (such as sodium) or a gas (such as helium) to prevent the energetic neutrons generated during the fission process from being slowed down.

Fertile—a material that becomes fissile upon absorbing a neutron.

Fissile—a material that will fission, i.e. split into two or more lighter materials, upon absorbing a neutron.

Fission—the process of splitting a heavy atom into two or more lighter atoms upon absorbing a neutron. This process generates a large amount of energy and usually at least two neutrons.

Fission products—elements that remain after nuclear fission has taken place.

Fossil fuel—carbon based fuel resulting from millions of years of biological decay. Coal, oil, and natural gas are the most common examples.

Fuel—materials that can be converted into useful energy (see nuclear fuel for the specific application to nuclear energy).

Fuel cycle—all the steps required to supply, use, and process fuel for nuclear reactors, including the disposal of nuclear wastes.

Fusion—fusing two light atoms into one heavier atom (a process that releases an enormous amount of energy).

Gamma ray—a high energy, short wavelength electromagnetic radiation emitted in the radioactive decay of many nuclides. Gamma rays are highly penetrating.

Geologic repository—a burial ground deep beneath the earth's surface designed for the long-term storage of high level nuclear waste.

Geothermal—an energy source arising from naturally occurring hot rocks beneath the surface of the earth. This is actually a form of nuclear energy, since the source of heat for the hot rocks is natural radioactive decay.

Greenhouse effect—heating of the earth's atmosphere due to the presence of certain gases (e.g. carbon dioxide) that trap energy from sunlight striking the surface of the earth.

Gross national product (GNP)—the total economic output of the nation in any given year (this includes national citizen output that is actually produced in foreign countries). Gross domestic product (GDP) is limited to productivity within the national borders.

Half-life—the length of time for a radioactive substance to lose half of its activity due to radioactive decay. At the end of one half-life, only 50% of the original radionuclide remains.

Heavy water—water in which the hydrogen atoms contain one neutron in their nucleus in addition to the characteristic proton.

High level waste (HLW)—highly radioactive solid material that results from the chemical reprocessing of spent nuclear fuel. It consists mainly of fission products, but also trace amounts of uranium and plutonium, plus other transuranic elements. NOTE: This definition could change with time as new uses of HLW become recognized and commercially employed (e.g. medical and industrial applications).

Hormesis—a term that describes biologically beneficial effects of low level radiation.

Hydroelectric—a process for producing electricity by utilizing water behind river dams to turn electrical generators.

IAEA—International Atomic Energy Agency, an agency created in 1954 within the United Nations to create and apply international safeguards consistent with promoting the peaceful uses of atomic energy while simultaneously preventing new nations from making nuclear weapons.

IFR—Integral Fast Reactor, a particular version of the liquid metal cooled fast reactor (LMR) based on an integral (or closed) fuel cycle at the site. A key feature of this concept, developed by the Argonne National Laboratory, is that the plutonium never occurs in a pure form. This provides potentially attractive nuclear proliferation resistant characteristics.

Ionizing radiation—radiation of sufficiently high energy that it can remove one or more electrons from the struck atom, thus leaving positively charged particles behind. High enough doses of ionizing radiation could cause cellular damage.

Isotopes—atoms of the same element but with different mass numbers. They contain the same number of protons in their nucleus (hence, the same chemical properties), but different numbers of neutrons, e.g. isotopes of carbon are C-12, C-13, and C-14.

Kilowatt-hour (kWh)—an energy unit defined as 1,000 watts of electricity for one hour (equivalent to 3,413 BTUs).

Licensing—giving a permit to build or operate a facility.

Linear hypotheses—the assumption that any radiation causes biological damage, according to a straight-line graph of adverse health effects versus dose.

LLE—Loss-of-Life-Expectancy, the average amount by which one's life is shortened by a particular risk.

LMR—Liquid Metal-cooled Reactor, a nuclear reactor in which the heat is removed by a liquid metal coolant (usually sodium). Since the highly energetic neutrons created during the fission process are not slowed down very much by this relatively heavy coolant, the neutrons remain at a fairly high speed. Hence, the LMR is often called a fast reactor.

Low level waste (LLW)—wastes from the nuclear industry containing very low amounts of radioactivity, requiring essentially no shielding or heat removal.

MHTGR—Modular High Temperature Gas Reactor, a nuclear reactor in which the heat is removed by a gas (usually helium). This reactor operates at higher coolant temperatures than other systems because the coolant is already in a gaseous form (negating concerns of coolant boiling).

mrem—millirem, a common unit of radiation dose defined as one-thousandth of a rem. An average American receives about 370 mrem of radiation in a year. Other units are also used to measure radiation.

MRS—Monitored Retrievable Storage, a facility in which spent nuclear fuel can be stored and monitored on a temporary basis.

Natural uranium—uranium as mined (containing 0.7% U-235 and 99.3% U-238).

Neutron—a basic particle residing in the nucleus of an atom that is electrically neutral. It has a mass approximately equal to a proton.

Nitrogen dioxide—a gas molecule consisting of one nitrogen atom and two oxygen atoms that is a combustion by-product of burning fossil fuels. The combustion temperature is high enough to oxidize atmospheric nitrogen.

Nonproliferation—a policy to discourage (and even prevent) non-nuclear weapons nations from acquiring nuclear bombs.

NPT—Nuclear Proliferation Treaty, an international treaty ratified in 1970 in which signatory nations agreed to submit to international safeguards (administered by the IAEA) to prevent the spread of nuclear weapons.

NRC—Nuclear Regulatory Commission, a U.S. agency chartered to develop and administer rules for regulating commercial nuclear applications (including nuclear power plants, medical and industrial uses).

NRDC—Natural Resources Defense Council, a private organization developed to champion "environmental" causes.

Nuclear energy—energy, usually in the form of heat or electricity, produced by the process of nuclear fission within a nuclear reactor. The coolant that removes the heat from the nuclear reactor is normally used to boil water, and the resultant steam drives steam turbines that rotate electrical generators.

Nuclear fuel—fissionable materials that can be "burned" (fissioned) in a nuclear reactor (e.g., uranium-235).

Nuclear proliferation—the process of new nations, not currently having nuclear bombs, acquiring nuclear warheads.

Nuclear waste—waste that is produced within the nuclear industry. This includes nuclear power plants, hospitals and medical laboratories, and numerous industrial users of nuclear products. See HLW, LLW, and TRU for definitions of specific types of nuclear waste.

Nucleus—the massive center of all atoms. It contains both protons (the number of which uniquely defines the element) and neutrons.

Nuclides—a general term used to describe the full range of elements and their family of isotopes.

Once-through—a fuel cycle in which spent fuel is not reprocessed.

Ozone—an atmospheric gas consisting of three atoms of oxygen in the molecule.

P-T—Partitioning-Transmutation: this refers to a process whereby spent nuclear fuel is chemically partitioned (i.e., separated into its constituent parts) and then converted (transmuted) into stable elements or isotopes by neutron bombardment.

Partitioning—the process of chemically separating spent nuclear fuel into its constituent parts (nuclides). It is akin to separating our household trash into baskets of glass, wood, paper, etc.

Periodic chart—a chart containing all the nuclides, i.e., all elements and their family of isotopes.

pH—a measure of acidity and alkalinity. A pH of 0 is maximum acidity (battery acid is about 1.0) and a pH of 14 is maximum alkalinity (lye is about 13.0). Neutral pH, neither acid or alkaline, is 7.0.

Plasma—a gas so hot that all electrons are stripped away from the atoms. As such, the gas is positively charged and can be confined in a magnetic field.

Plutonium—a very heavy element formed by neutron absorption in uranium-238. Plutonium, like uranium, has two principal isotopes that are fissile.

Primary system—the enclosed coolant system within a nuclear reactor in which heat is directly removed from the reactor.

Proton—a basic particle residing in the nucleus of an atom that is positively charged. It has a mass approximately equal to a neutron.

Quad—a measure of energy equivalent to a quadrillion (a million times a million times one thousand, or 10^{15}) British thermal units (BTU).

Radiation—particles or rays (waves) coming from radioactive substances.

Radioactivity—spontaneous emission of radiation from the unstable nucleus of an atom.

Radioisotope—a radioactive isotope.

Radionuclide—any species of an atom that is radioactive. A generic word used to replace radioisotope, which is limited to one element.

Radiotoxicity—the toxicity to human cells caused by absorption of high doses of radioactive substances. Many chemicals are toxic or poisons at high doses

Radon—a natural radioactive gas produced by the radioactive decay of radium, a decay product of uranium.

Reactor—a nuclear device involving neutrons to induce a fission chain reaction.

Recycling—a term made popular in the environmental movement to reuse materials that otherwise would be discarded as waste. Within the nuclear industry, it is a synonym for reprocessing spent fuel.

Regulation—maintenance of standards of performance through rules.

Rem—(Roentgen equivalent man), a unit used in radiation protection to measure the amount of damage to human tissue from a dose of ionizing radiation. An average American receives about 0.370 rems of radiation per year.

Reprocessing—the mechanical and chemical processing of spent nuclear fuel to separate useable products (i.e., uranium and plutonium) from waste material (i.e., fission products). As time progresses, much of what is currently defined as "waste" will likely find commercial applications.

Secondary system—the enclosed coolant system within a nuclear power plant in which heat is transferred from the primary system to the steam generators. Not all nuclear power plants require a secondary system.

Shale—compacted clay rock.

Shipping cask—a container that provides appropriate shielding and structural rigidity for the transportation of radioactive material.

Solar energy—energy that is derived from our sun. Strictly speaking, direct solar radiation, hydroelectric, wind, and even biomass and fossil energy are all forms of solar energy since none would be present without the existence of the sun. However, for this book, only direct solar radiation is denoted by this term.

Spent fuel—nuclear fuel elements that are discharged from a nuclear reactor after they have been used to produce power.

Sulfur dioxide—a gas molecule consisting of one sulfur atom and two oxygen atoms that is produced as a by-product of burning fossil fuels that contain sulfur impurities.

Sustainable development—a term used to connote ways humanity must learn to use and develop its resources in order to sustain a high quality of life on the planet Earth for periods well into the future.

Thermal reactors—nuclear reactors that employ very slow ("thermalized") neutrons to sustain the fission process. Water is commonly used in such reactors for a coolant since the hydrogen contained in water is very effective in slowing down the highly energetic neutrons generated during fission.

Toxicity—a measure of significant biological damage done by the absorption of a foreign substance. Many toxic materials, including radiation, are not toxic at low levels.

Transmutation—the process of changing one isotope into another using nuclear reactions.

Transuranics (TRU)—nuclides with an atomic number greater than uranium (i.e., greater than 92). The principal transuranics are neptunium (No. 93), plutonium (No. 94), and americium (No. 95).

Uranium—the heaviest element normally found in nature. The fissile isotope uranium-235 is the principal nuclear fuel material used in today's nuclear power reactors.

Vitrification—the process of placing nuclear high-level waste (HLW) into a glass form for long-term disposal.

X-rays—electromagnetic radiation of energy greater than that of visible light, usually produced by an x-ray machine. Physically, x-rays and gamma rays are similar.

NOTE: The author is indebted to the following references for much of the material used for these definitions:

1. Murray, Raymond L., *Understanding Radioactive Waste*, Fourth Edition, 1994, Battelle Press, 505 King Avenue, Columbus, Ohio 43201.

2. *The Nuclear Waste Primer*, The League of Women Voters, 1993, Revised Edition, OCRWM Information Center, P.O. Box 44375, Washington, D.C. 20026.

NOTES

Chapter 1: So What's the Big Fuss All about?

1. World population, *The World Almanac & Book of Facts 1995*, Funk & Wagnells, p. 839 and *1981 Statistical Yearbook*, 32th Edition, United Nations, p. 2.; World energy consumption, *The World Almanac & Book of Facts 1995*, ibid, p. 164, and Robert L. Loftness, *Energy Handbook*, Van Nostrand Reinhold Co., 1978, p. 121.

2. The United States is now beginning to use the term Gross Domestic Product (GDP), to focus on goods and services produced within the geographic boundaries of the United States (excluding foreign gains by U.S. citizens). However, I will continue to use GNP because of the extensive amount of information available, both in the U.S. and worldwide.

3. Compiled from *Statistical Abstract of the United States*, 112th Edition, Bernan Press, 1992, Table 1371 (p. 831), and Table 1397 (p. 844, 845). Data are for 1989.

4. U.S. population, *The World Almanac & Book of Facts 1995*, Funk & Wagnells, p. 376, 377; U.S. energy consumption, Robert L. Loftness, *Energy Handbook*, Van Nostrand Reinhold Co., 1978, p. 138, and *Annual Energy Review 1991*, Energy Information Administration, Table 5 (p. 15).

5. Energy consumption from *Annual Energy Review 1991*, Energy Information Administration, June 1992, Table 5 (p. 15); GNP from *Survey of Current Business*, Feb. 1992, U.S. Department of Commerce, Vol. 72, No. 2, Table 2 (p. 32).

6. Data from reference 5 above, years 1974-1990.

7. Maxey, Margaret N., "Nuclear Energy Politics: Moralism vs. Ethics," talk given to the Edison Electric Institute 45th Convention, 1977.

8. GNP data from *Statistical Abstract of the United States*, 112th Edition, Bernan Press, 1992, Table 1371 (p. 831); Income spent on food, ibid, Table 1376 (p. 833); Life expectancy and infant mortality, ibid, Table 1361 (p. 824); Literacy from *Information Please ALMANAC, 1995*, 48th Edition, Houghton Mifflin Co., pp. 202, 287; Energy production, ibid, pp. 139, 140; India population from *The Europa World Yearbook*, Vol. 1, 1994, p. 1429; U.S. population from *The World Almanac & Book of Facts 1995*, Funk & Wagnalls, p. 375.

9. Hafele, Wolf, *Energy in a Finite World; Paths to a Sustainable Future*, Ballinger Publishing Company, Cambridge, Mass, 1981.

10. Starr, Chauncey and Milton F. Searl, "Global Electricity Future: Demand and Supply Alternatives," Electric Power Research Institute, Palo Alto, CA, Nov. 27, 1989.

11. Biomass refers to burning trees or other biological materials. See chapter 3.

12. Bloomster, C. H. and E. T. Merrill, "Potential Growth of Nuclear and Coal Electricity Generation in the U.S.," PNL-6977, Pacific Northwest Laboratory, Richland, WA, 99352, August, 1989.

13. The basic study was conducted in support of DOE's national energy strategy (cf. U.S. Department of Energy, *National Energy Strategy*, First Edition, 1991/1992, p. 108). The lowest predicted need for new electrical capacity in the 1991 to 2010 study period was 190 gigawatts, i.e., 190 plants of 1,000 megawatt capacity.

Chapter 2: But What about the Things I Hear?

1. Three German scientists, Otto Hahn, Fritz Strassman, and Lise Meitner discovered in 1939 that certain atoms of uranium would literally split apart (fission) when bombarded with neutrons. This fission process releases a large amount of energy.

2. Rothman, S. and Lichter, S.R., "The Nuclear Energy Debate: Scientists, the Media, and the Public," *Public Opinion* August 1982, p. 47.

3. Bisconti, Ann S., "Gallup Survey: More and More People Favor Nuclear Energy," *Public Opinion*, Nuclear Energy Institute, May 1994, p.2.

Chapter 3: Choices, Choices, Choices

1. Oil & Gas Journal, Reported in *Annual Energy Review 1990*, Energy Information Administration, Table 113, p. 257. (Only the largest petroleum nations are included in the figure.)

2. This tabulation attributes nearly half of the global coal reserves to mainland China, although British Petroleum (in the same reference) credits China with only about 15% of the world's supply. The estimates for all other countries are in good agreement.

3. Synfuel is a term used to describe a liquid fuel made from coal.

4. World Energy Council, 1989 Survey of Energy News, Reported in *Annual Energy Review 1991*, Energy Information Administration. (Only the largest coal-producing nations are included in the figure.)

5. The photovoltaic process allows solar energy to be converted directly into electricity.

6. Inhaber, Herbert, "Risks with Energy from Conventional and Nonconventional Sources," *Science* 203, 1979, 718-723.

7. Teller, Edward, *Energy From Heaven and Earth*, W.H. Freeman & Company, San Francisco, 1979, p. 239.

8. Ordinary water is composed of two hydrogen atoms and one oxygen atom (H_2O). In the rare case that the hydrogen atoms contain a neutron, in addition to a proton, it is called deuterium. Two atoms of deuterium combined with one atom of oxygen is called "heavy water." Tritium is a hydrogen atom containing two neutrons plus the one proton.

9. There have been about a half dozen deaths within the military during the early years of nuclear energy. Furthermore, it is possible that small amounts of radiation released during isolated mishaps could result in one or more deaths in the United States due to latent cancer. However, as we shall see in chapters 4 and 5, the probability of such deaths is essentially zero.

10. United States Council on Energy Awareness (USCEA) *Energy Data*, (Suite 400, 1776 I Street, N.W., Washington, D.C., 20006) September 1991, p. 40.

11. A "fertile" material is a substance that, upon absorbing a neutron, is transmuted (transformed) into a "fissile" material. When the latter material absorbs a neutron, it will split apart (fission), releasing a large amount of energy in the process.

12. The word "fast" in fast reactor refers to the relative speed of the neutrons doing the work inside the reactor. In the common "thermal" power reactors, neutrons generated during the fission process are intentionally slowed

down (thermalized) by a "moderator" such as water or graphite to improve the probability of sustaining the fission process. If the intention is to convert fertile material into fissile material, however, it is highly advantageous to allow the neutrons to maintain the high (fast) speeds of their birth. This requires the avoidance of water as a coolant. Liquid sodium is a popular choice.

13. Uranium availability determined as Reserves plus Probable from Alan E. Waltar and Albert B. Reynolds, *Fast Breeder Reactors*, Pergamon Press, 1991, Table 1-5; Oil and natural gas availability determined from reference 1; Coal availability determined from reference 4.

14. Data for 1850-1950 developed from Fred H. Schmidt and David Bodansky, *The Fight over Nuclear Power*, Albion Publishing Co., 1736 Stockton Street, San Francisco, CA, 94133, 1976 (p. 12); Data for 1950-1990 taken from *Annual Energy Review 1991*, Energy Information Administration, June 1992, Table 3 (p. 11).

CHAPTER 4: Radiation—The Invisible Enigma

1. Since our total yearly average radiation exposure is about 365 mrem, Dr. John Cameron has suggested that we simply interpret the numbers represented in figure 17 as the number of days of Background Equivalent Radiation Time (BERT). For instance, under the term "elective," we could say that choosing to fly from coast-to-coast would subject us to 2 BERTs, i.e. 2 days of our annual radiation background.

2. Strictly speaking, radon does not contribute any radiation directly. It is the daughter products of radon (species in its radioactive decay chain), which are solids and attach to dust particles, that provide the source of our "radon" dose.

3. Cohen, Bernard L., *Before It's Too Late*, Plenum Publishing Corp., 233 Spring Street, New York, 10013, 1983. See also Cohen, Bernard L., *The Nuclear Energy Option*, Plenum Publishing Corp., 233 Spring Street, New York, 10013, 1990.

4. Eisenbud, Merril, *Environmental Radioactivity from Natural, Industrial, and Military Sources*, Orlando: Academic Press, 1987 (3rd Edition).

5. Cameron, John, "What Does the Nuclear Shipyard Worker Study Tell Us?" American Nuclear Society Winter Meeting, Vol. 71, November 13, 1994, p. 36.

6. Evans, Robley, "Radiation in Man," *Health Physics*, Nov. 1974, pp. 497-510.

7. The term "hormesis" refers to instances where stimulation of human body cells by external forces improves overall health and vitality.

8. Luckey, T. D., *Hormesis With Ionizing Radiation*, CRC Press, Inc., 2000 NW 24th Street, Boca Raton, FL, 33431, 1980.

9. Lee, Robert W., "Why Nuclear Power?" *The New American*, March 27, 1989, 31-41.

CHAPTER 5: After Chernobyl? You've Got to Be Kidding!

1. I'm looking at this question from the standpoint of a skeptical member of the public who is concerned about health hazards. From the vantage point

of many members of the nuclear industry, particularly the utilities who have to obtain the operating license, many would say the process is hopelessly bogged down. They believe that the NRC is far too strict, and requires levels of safety well in excess of what is realistically necessary. They have some validity to their arguments, since the expenses of continual safety upgrades eventually get passed on to the consumer, and the statutory charter of the utility is to deliver electricity to the public at the lowest cost possible.

2. duPont, Robert L., *Nuclear Phobia—Phobic Thinking About Nuclear Power*, Media Institute, 1627 K Street NW, Suite 201, Washington, DC, 20006, 1980.

3. None of the early Soviet reactors, including the Chernobyl type (RBMK) reactors, were equipped with containment structures. However, the newest Russian reactors (the VVER-1000) designs and some of the newest VVER-440 designs) do have Western-type containment buildings.

4. Lee, Robert W., "Why Nuclear Power?" *The New American*, March 27, 1989, 31-41.

5. International Atomic Energy Agency (IAEA), *The International Chernobyl Project: An Overview*, 92-0-129091-8, IAEA, Vienna, Austria, 1991.

6. Those organizations include the Commission of the European Communities (CEC), the Food and Agriculture Organization of the United Nations (FAO), the International Atomic Energy Agency (IAEA), the International Labour Office (ILO), the United Nations Scientific Committee on the Effects of Atomic Radiation (UNSCEAR), the World Health Organization (WHO), and the World Meteorological Organization (WMO).

7. Since completion of the study, there have been reported increases in thyroid tumors in young children. Considerable controversy exists regarding the causes of such increases. Certainly this observation is of concern, and demands special medical attention for the victims. However, the individual health risks for children are still low, equivalent to the normal adult risk of contracting thyroid tumors.

8. It may be helpful to note that of the approximately 100,000 Japanese atomic bomb survivors in Hiroshima and Nagasaki, the total number of increased cancer deaths is about 500, which is less than would be expected in an unexposed population of this size. This population was subjected to a substantially higher dose than Chernobyl victims. Ironically, the remaining Japanese survivors have longer average life spans than the unexposed populations.

CHAPTER 6: Not in My Backyard

1. Murray, Raymond, *Understanding Radioactive Waste*, Third Edition, Battelle Press, Columbus, Ohio. p 87.

2. It is likely that the definition of high-level waste (HLW) will change with time as new beneficial uses for various HLW constituents become commercialized.

3. Lee, Robert W., "Why Nuclear Power?" *The New American*, March 27, 1989, 31-41.

4. Schmidt, Fred H. and David Bodansky, *The Energy Controversy: The Fight Over Nuclear Power*, Albion Publishing Company, 1736 Stockton Street, San Francisco, CA, 94133, 1976.

5. It is true that the high-level waste material that we must deal with can be somewhat complicated chemically. Also, it generates a considerable amount of heat that must be dissipated for a long period of time. However, there is little question of developing technically acceptable glass schemes.

6. Russ, George D., Jr., *Nuclear Waste Disposal: Closing the Circle*, Atomic Industrial Forum, Inc., pp. 17-18.

7. Cohen, Bernard L., *Before It's Too Late*, Plenum Publishing Corp., 233 Spring Street, New York, N.Y., 10013, 1983.

8. Cowan, George A., "A Natural Fission Reactor," *Scientific American* 235, July 1976, pp. 36-47.

9. Some advanced fuel cycle schemes have been proposed that would accomplish essentially all of the reprocessing steps on-site, such that cross-country transportation would be minimized or even eliminated.

10. Murray, Raymond, *Understanding Radioactive Work*, ibid, p. 91.

11. Cohen, Bernard L., "Ocean Dumping of High-Level Waste—an Acceptable Solution We Can 'Guarantee,'" *Nuclear Technology* 47, 163, January 1980.

12. Hollister, Anderson, and Heath, "Subseabed Disposal of Nuclear Wastes," *Science* 213, 1321, September 18, 1981.

13. The latent energy content in spent fuel discharged from 30 years of operations in a present light water reactor, if recycled and used in breeder reactors, is equivalent to about 11 billion barrels of oil (2 years of current U.S. oil usage).

CHAPTER 7: No More Bombs, Please

1. The nations that have declared their possession of nuclear weapons are the United States, Russia, Britain, France, China, and India. There continue to be rumors that a few other nations, such as Pakistan, Israel, and North Korea are nearing or possibly have such capability. No confirmation of such status has been obtained to date. South Africa has voluntarily given up its status as a nuclear weapons nation. The former Soviet bloc nations of Ukraine, Byelorussia, and Kahsakstan have declared their intentions to relinquish their roles in the nuclear weapons enterprise of Russia.

2. Clancy, Tom, *The Sum of All Fears*, Berkeley Books, The Berkeley Publishing Group, 1992.

CHAPTER 8: I Don't Want to Take Chances

1. Cohen, Bernard L., *Before It's Too Late*, Plenum Publishing Corp., 233 Spring Street, New York, N.Y., 10013, 1983.

2. Cohen, B. L. and I. S. Lee, "A Catalog of Risks," *Health Physics* 36, 707 (1979); See also the update, Bernard L. Cohen, "Catalog of Risks Extended and Updated," *Health Physics* 61, No. 3, September 1991, pp 317-335.

3. To calculate a typical loss-of-life expectancy (LLE), suppose a 40-year-old person takes a risk that has a 1% chance of being immediately fatal. Actuarial tables show that this person has a life expectancy of another 37.3 years (i.e., a normal life expectancy of 77.3 years) if this risk were not incurred. Consequently, the LLE for this activity is 0.373 years (0.01 x 37.3 years). This does not mean that this person will die 0.373 years sooner as a result of taking this risk. Rather, it means that if 100 people were to take this same risk, one would die immediately, having his or her life shortened by 37.3 years, and the other 99 would not have their lives shortened at all. Nevertheless, the weighted LLE of 0.373 years represents a meaningful and tangible way to grasp the magnitude of the risk involved.

4. Ames, Bruce and Lois Gold, *Phantom Risk: Scientific Inference and the Law*, MIT Press, Cambridge, MA, 1993.

5. By considering the analogy attributed to MIT Professor Rasmussen in chapter 5, it is easy to see how an intentionally critical mind could speculate an essentially unbounded level of concern. It is safe to say that well over 99% of the nuclear professionals in this country would seriously doubt the extreme UCS numbers.

6. duPont, Robert L., *Nuclear Phobia—Phobic Thinking About Nuclear Power*, Media Institute, 1627 K Street NW, Suite 201, Washington, DC, 20006, 1980.

7. Fortunately, some evidence is becoming available that indicates public objections to nuclear energy may not be phobic; rather, it may be that such objections can be constructively addressed. Dr. Peter Sandman, a noted risk analyst from Rutgers University, has compiled impressive evidence that the elements identified by Dr. duPont, plus several other elements, can be quantified and dealt with. Dr. Sandman calls such elements key ingredients of an "outrage" factor, which must be added to the "hazards" factor that scientists define as probability of an event times the consequences of that event. Elements such as voluntary (vs involuntary), control (vs lack of control), familiar (vs unfamiliar) are all essential ingredients. It is likely that the nuclear industry has often failed to recognize the implications of such elements and, therefore, has not dealt with them in an appropriate manner. See Sandman, Peter M., "Risk Communications: Facing Public Outrage," *EPA Journal*, November 1987, pp 21-22.

8. Wilson, Richard, "Analyzing the Daily Risks of Life," *Technology Review*, February 1979, pp. 41-46.

9. It might be argued that some of these small numbers cannot be determined with the accuracy indicated. In fact, for the "low doses" of cigarettes, peanut butter, etc., the actual health effects may be lower than shown. That is because these numbers were derived using the same linear projections as employed for determining the health effects of low-level radiation. The human body may be able to handle low doses of many substances without detectable harm.

10. Allman, William F., "We Have Nothing to Fear—But a Few Zillion Things," *Science*, October 1985, pp. 38-41.

11. Breyer, Steven, *Breaking the Vicious Circle: Toward Effective Risk Regulation*, Harvard University Press, 1993, Cambridge, MA.

CHAPTER 9: But I'm an Environmentalist

1. Eisenbud, Merrill, *How Clean is Clean? How Safe is Safe?*, Cogito Books, 4513 Vernon Blvd., Madison, WI, 53705, 1993.

2. Anyone interested in gaining a clear understanding of the environmental movement in the U.S., including the deceptive mixture of technology and politics that has transpired, would benefit enormously from reading Dr. Dixy Lee Ray's insightful books, *Trashing the Planet* and *Environmental Overkill*. In these well-documented and very readable books, Dr. Ray cites several examples of how special interest groups have successfully thwarted good science and left tragic public consequences in their wake, all in the name of the environment. The present accounts on DDT and Alar are summarized from her excellent works. See Ray, Dixy Lee, *Trashing the Planet*, Regnery Gateway, 1130 17th Street, NW, Washington, DC, 20036, 1990, and Ray, Dixy Lee, *Environmental Overkill*, Regnery Gateway, 1130 17th Street, NW, Washington, DC, 20036, 1993.

3. Carson, Rachel, *Silent Spring*, Houghton Mifflin, Boston, 1962.

4. Huber, Peter W., *Galileo's Revenge: Junk Science in the Courtroom*, BasicBooks, A division of Harper Collins Publishers, 1991.

5. *Science*, "Global Temperature Hits Record Again," Vol. 251, January 18, 1991.

6. Ray, Dixy Lee, *Trashing the Planet*, Regnery Gateway, 1130 17th Street, NW, Washington, DC, 20036, 1990.

7. Michaels, Patrick, "Acid Rain vs. the Greenhouse Effect," *World Climate Review*, Vol. 2, No. 3, Spring 1994, p. 9.

8. Abrahamson, Dean Edwin, *The Challenge of Global Warming*, Island Press, Suite 300, 1718 Connecticut Avenue NW, Washington, DC, 20009, 1989.

9. *Power Engineering*, "French CO_2 Emissions Fall as Nuclear Output Rises," Penwell Publishing Company, June 1989.

10. Legault, Benoit, "The Environmental Challenge—Mother Nature Calls for Help," *Ascent*, Vol. 8, No. 1, Spring 1989, Atomic Energy of Canada Limited, 344 Slater Street, Ottawa, Ontario, K1A OS4.

11. Minister of Supply & Services Canada, *The Acid Rain Story*, Catalog No. En 21-40, 1984, Environment Canada, Ottawa, Ontario, K1A OH3.

12. Burge, Ray, "The Environmental Case for Nuclear Power," *Ascent*, Vol. 8, No. 1, Spring 1989, Atomic Energy of Canada Limited, 344 Slater Street, Ottawa, Ontario, K1A OS4.

13. Science Concepts, Inc., "Reducing Airborne Emission with Nuclear Electricity," for U.S. Council for Energy Awareness, December 1989, 1625 K Street NW, Suite 1125, Washington, DC, 20006.

14. Michael R. Fox, private communication, March 28, 1995.

15. Cohen, Bernard L., *The Nuclear Energy Option*, Plenum Corp., 233 Spring Street, New York, N.Y., 10013, 1990, p. 246.

16. Cohen, Bernard L., *Before It's Too Late*, Plenum Publishing Corp., 233 Spring Street, New York, N.Y., 10013, 1983.

CHAPTER 10: I'll Wait and See

1. Maxey, Margaret N., "Nuclear Energy Politics: Moralism vs. Ethics," talk given to the Edison Electric Institute 45th Convention, 1977.

2. *Isotopes in Everyday Life*, International Atomic Energy Agency, Vienna, Austria, December 1990; Excerpts presented by Richard T. Kennedy, "The Atom and Human Values," *American Nuclear Society Transactions*, Vol. 65, Boston, MA, June 10, 1992;

3. "The Untold Story: Economic and Environmental Benefits of the Use of Radioisotopes and Radioactive Materials," Management Information Services, Inc., October 1993 (report prepared for Organizations United for Responsible Low-level Radioactive Waste Solutions).

4. "Economic and Employment Benefits of the Use of Nuclear Energy to Produce Electricity," Management Information Services, Inc., February 1994 (report prepared for the U.S. Council for Energy Awareness).

5. Cook, James, "Nuclear Follies," *Forbes Magazine*, February 11, 1985, p.82.

6. Moynet, G., "Electricity Generation Cost Assessments Made in 1987 for Stations to be Commissioned in 1995," UNIPEDE Conference, Sorrento, Italy, May 24–June 3, 1988.

7. Hubbert, M. King, "The World's Evolving Energy System," *American Journal of Physics*, Vol 49, No. 11, November 1981 (with updates from Twentieth Century Petroleum Statistics, 1992, DeGolyer & MacNaughton, One Energy Square, Dallas, Texas, 75206).

Index

---

* Page numbers in bold indicate the word's location in the glossary.

Nuclear Energy, Radiation and Health

My Life With Radiation
Hiroshima Plus Fifty Years
By Ralph E. Lapp
1995, 141 pages, softcover
with b&w photos; glossary

This is the story of Dr. Ralph Lapp's extensive involvement with radiation and nuclear bombs. Concerned about the dangers of military secrecy, he changed his career from measuring radiation at nuclear testing sites for the military to spreading his information to the public through lectures, books, and articles. A combination of personal reflection and scientific analysis.

How Clean is Clean?
How Safe is Safe?
A Review of Environmental Priorities
By Merril Eisenbud
1993, 63 pages, softcover

"Although it is a slim volume, Eisenbud's book is a heavy hitter. It should be required reading for Congressmen and high school teachers . . ." Ralph Lapp, *Health Physics*

Written by a pioneer environmentalist, this book traces the evolution of the modern environmental movement. The author is currently Professor Emeritus of Environmental Medicine at New York University Medical Center and has been a consultant to the United Nations, private industry, and to the U.S. and foreign governments.

Understanding Radiation
By Bjorn Wahlstrom
1996, 120 pages, softcover
with 2-color illustrations; Index

An expert in the field of radiation explains the subject for the layperson. He uses the language of the non-scientist to explain what radiation is, where it comes from, how it is used and how it is measured. He also explains the health risks involved in industrial radiation uses, including nuclear power.

The Good News About Radiation
By John Lenihan
1995, 173 pages, softcover
Appendices; Index

". . . a rare book—one that emphasizes the benefits of radiation exposure . . . a well-written introduction to the risks and benefits of radiation for the lay person. It is easy to read yet it does not in general mistreat the technicalities of the subject. It is stimulating and provocative. I recommend it." Radiological Protection Bulletin, 1995

This is a cheerful book with a serious purpose. It discusses the role of radiation in nature, how it increases the efficiency of industry as well as how it may actually improve your health!

Call or write for our free catalog containing complete pricing and ordering information on these and other science books for the general public.

Cogito Books

4513 Vernon Boulevard
Madison, WI 53705-4964

800-442-5778 • FAX: 608-265-2121 • E-MAIL: MPP@MACC.WISC.EDU

Cogito Books is an imprint of Medical Physics Publishing,
a nonprofit, tax-exempt organization formed for educational and scientific purposes.

About the Author

Alan E. Waltar is Past President of the American Nuclear Society (ANS), the largest professional organization of nuclear scientists, engineers, and educators in the world. He is the Founding President of the Eagle Alliance, now gaining momentum as a major new educational movement of national laboratories, universities, corporations, professional associations, unions, and concerned individuals dedicated to the revitalization of nuclear science and technology in America.

Dr. Waltar obtained Bachelor, Master, and Doctorate degrees from the University of Washington (Seattle), M.I.T. (Cambridge), and the University of California (Berkeley), respectively. He has led numerous engineering projects and held several management positions at Westinghouse Hanford Company. As an outgrowth of his teaching experience at the University of Virginia Department of Nuclear Engineering, Dr. Waltar wrote (with Professor A.B. Reynolds) a textbook on fast breeder reactors that, with the Russian translation edition, has become the standard international text in the field. In addition to organizing numerous international technical conferences, Dr. Waltar has published over 50 refereed scientific articles. In 1984 he was elected Fellow of the ANS, the highest peer recognition for technical excellence in the Society. He currently serves as Director of International Programs for Advanced Nuclear and Medical Systems.

On the lighter side, Alan is a popular and familiar lead in Richland Light Opera Company productions. In his passion to achieve public understanding of energy issues, he has given scores of lectures to service groups, colleges, churches, and scientific organizations throughout the United States and 15 foreign countries. Several years ago, Alan and his wife, Anna, cofounded a non-profit corporation to build a retreat center in the natural setting of the Central Washington mountains. It is largely through this experience that Alan developed such a keen sense for protecting our inherited environment for generations yet to come. Alan resides with his wife in Richland, Washington, where they raised four children now out of the nest.